P9-CEU-534

WINDOW INTO AN EGG

WINDOW INTO AN EGG

SEEING LIFE BEGIN

by Geraldine Lux Flanagan

WITH PHOTOGRAPHS

YOUNG SCOTT BOOKS

YOUNG SCOTT BOOKS
A DIVISION OF ADDISON-WESLEY PUBLISHING COMPANY, INC.
READING, MASSACHUSETTS 01867
LIBRARY OF CONGRESS CATALOG CARD NUMBER: 66-11411
ISBN: 0-201-09405-3
PRINTED IN THE UNITED STATES OF AMERICA
DESIGNED BY EVE METZ

SOURCES OF PHOTOGRAPHS

PAGES 8, 11 *bottom*, 14, 20-23, 27, 30, 35, 36, 39, 43, 44, 47, 48, 51, 53, 56-58: Mr. S. A. Buckingham

PAGES 38, 54, 62-68, 70, 71: Mr. Robert P. Matthews

PAGES 37, 50, 60, 61: courtesy of Professor Edward A. Shano, Cornell University

PAGES 31, 33, 34, 41: courtesy of Dr. Howard L. Hamilton; from *Lillie's Development of the Chick*, revised by Howard L. Hamilton, Holt, Rinehart & Winston, Inc., New York, 1952

PAGES 15 *bottom* and 16: courtesy of Dr. Marlow W. Olsen, U.S. Dept. of Agriculture, Agricultural Research Service, Animal Husbandry Division, Beltsville, Maryland

PAGES 24, 25: courtesy of Dr. Marlow W. Olsen; from *Journal of Morphology*, 70:533; figs. 16, 17, 22, 25

PAGE 10: Mr. L. Hugh Newman; from the Natural History Photographic Agency

PAGE 11 *top*: Radio Times Hulton Picture Library

PAGE 12: courtesy of Dr. C. Polge, Agricultural Research Council, Reproductive Physiology Research Centre, Cambridge, England

PAGE 13: courtesy of Dr. Thomas H. Clewe, Vanderbilt University School of Medicine, Department of Obstetrics and Gynecology

PAGE 15 *top*: courtesy of Dr. P. E. Lake, Agricultural Research Council, Poultry Research Centre, Edinburgh, Scotland

PAGE 17: courtesy of Dr. John J. T. Owen, Oxford University; from *Chromosoma*, *16*:601-608, Berlin-Heidelberg-New York: Springer, 1965

PAGE 29: courtesy of Professor Bradley M. Patten; from *Early Embryology of the Chick* by Bradley Patten, fourth edition, copyright 1951, McGraw-Hill Book Company, New York, used by permission of the publisher

PAGE 46: the drawing is by Miss Christine Court of the Department of Human Anatomy, Oxford University, England

JACKET/COVER: Mr. Brian Archer

To John and Cara and Christopher

I would like to convey my gratitude to all who so kindly helped with this book. The majority of photographs were taken by S. A. Buckingham and Robert P. Matthews. For additional photographs I am deeply indebted to Professor Bradley M. Patten, to Dr. Marlow W. Olsen, to Dr. Howard L. Hamilton, to Dr. Thomas H. Clewe, and to Dr. P. E. Lake. I particularly want to mention Professor Edward A. Shano for his most valuable advice and photographs; Dr. John J. T. Owen for his generous personal help; and Dr. C. Polge for providing the mammalian eggs shown on page 11.

In writing the book I greatly appreciated the kindness of Dr. Robert L. DeHaan in giving many vital corrections and suggestions. Among expert advisors, I was lucky to have a special one on the child's point of view: my son, John, whose thoughtful comments, assistance and interest helped to make work a pleasure.

Contents

This egg is filled with yolk and egg white. Yet, in three weeks a chick will hatch from it. How does the chick grow?

What Is an Egg?

All the animals and plants you can see around you began from an egg. Even you did. Have you ever thought about an egg, perhaps the one you know best, the chicken egg? It does not look alive, but it can turn into something very much alive. An ordinary chicken egg, which seems to have just a yolk and some egg white in it, can turn into a lively chick that comes hopping out of the shell.

You can think of an egg as a package. It is nature's neat and handy package that holds a new life from a mother and father. The mother and father may be people, or they may be a hen and rooster, or a male and female elephant, or goldfish, or grasshoppers. Trees, flowers, vegetables, all have eggs and their eggs turn into seeds that grow into new plants.

Inside every egg there is a very small living part that comes from the parents. It is just a dot, smaller than the dot on this "i". This dot can be alive and can grow, as you shall see. To grow, it will need food. Some eggs have a lot of food packed around the living dot. This makes them large eggs. The chicken egg is one of these large eggs, filled up with food. It is one of the largest eggs in the world. The largest of all contains even more food; it is the egg of another bird, the ostrich. That egg is as big as a grapefruit and weighs as much as a filled quart bottle of milk.

All birds' eggs, even the eggs of very small birds like hummingbirds, are well enough provided with food to be called big eggs, at least big enough to hold in your hand and look at. The same is true of the eggs of lizards, turtles, and snakes, and all the animals that are related to them. These families of animals lay eggs with enough stored-up food to last for many days. It takes the young of all these animals that much time to become ready to hatch.

There are other families of animals that also lay eggs. Their eggs are smaller, although still large enough to be plainly seen if you happen to find them. Perhaps you know some of them. These are the eggs of frogs, fish, insects, and all their relatives. The young of all these animals do not stay long inside the egg. They come out of the egg in a hurry, in only a few days, and can go out to get their own food right away. All they need in the egg is a snack. And that is all they get. Mother animals who produce so little food in their eggs, who lay the eggs and then leave them, in a way make up for this lack of motherly care by laying huge numbers of eggs. One female frog can lay several

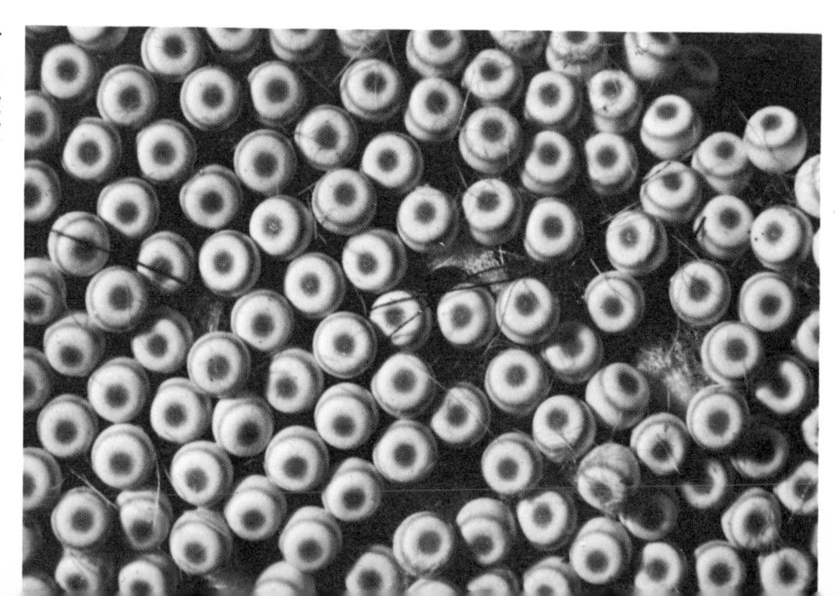

Insect eggs from Vapourer moth, five times real size. Living part and surrounding food are concealed by hard outer cover.

Fish eggs from trout, about twice real size, under water. Living part is in dark spot visible through transparent, jelly-like outer cover.

thousand eggs at one time, and some fish lay as many as twenty thousand! Most of these unprotected young fail to grow, or are gobbled up by enemies. But out of such enormous numbers of eggs, we can expect enough lucky survivors each year to give us plenty of new fish and frogs.

There is yet another kind of egg, so small that the whole egg is about as small as the tiny living dot in the hen's egg. These eggs are hardly visible, and to see them well you must look at them under a microscope. Small as they are, they are true eggs. Each is shaped like a tiny bubble that contains the living dot and a droplet of yellow yolk. These eggs are so small because they have almost no food inside them.

What animals could come from such tiny eggs? Well . . . animals like elephants, bears, whales, pigs, cows, puppies, and kittens, come from such small eggs. One of the smallest animals that comes from that kind of egg is a mouse.

You can never tell by the size of the egg how big an animal will come from it. The size of the egg only tells you how much

Three eggs from a pig are like dots. They are photographed slightly larger than true size on a glass slide.

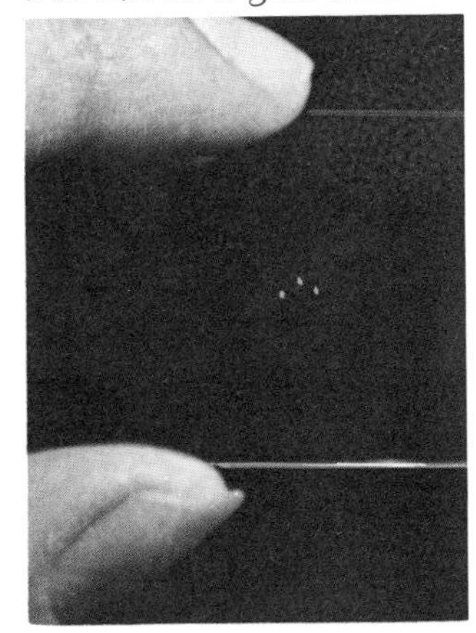

food the egg contains. Some of the biggest animals come from the smallest eggs. An elephant comes from a much, much smaller egg than a chick. The elephant's egg is even much smaller than the egg of a hummingbird. How can this be? Where does the elephant get food to grow from the tiny living dot and become big enough to be born?

This gives you a clue to the answer: the elephant is born, it does not hatch from the egg like a chick. It grows from egg to elephant inside the elephant mother's body and is nourished there. The elephant, and the other animals that grow from the nearly invisible eggs, belong to an animal family called mammals.

Mammal mothers never lay their eggs; their young always grow and are nourished within the mother's body until they are ready to be born. That is why mammal eggs can be so small, they do not need the packed-up food. *You* have such a mother, and that is why you have come from an egg that was so small. A mouse, and an elephant, and you, come from an egg of about the same size. The elephant grows big inside its big mother for close to two years; the mouse grows inside its little mother for only three weeks. You grew for nine months, before you were born.

Pig egg seen through microscope and magnified about 250 times. Within transparent, soft shell is the darker-looking living part with minute amounts of food.

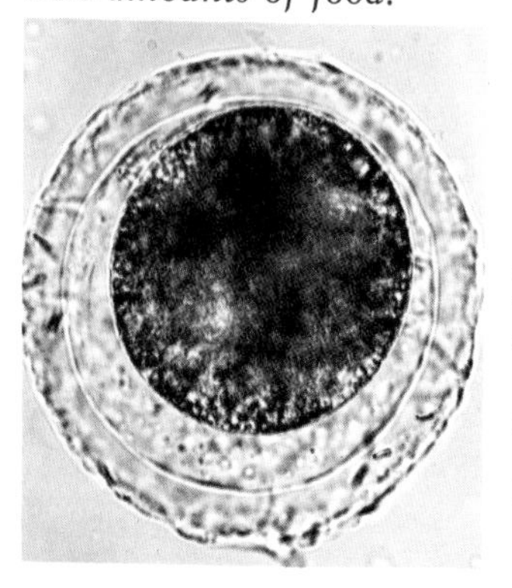

Perhaps you think it sounds silly to say you have once been an egg. It may be all right for chicks, but not for you. Therefore, you might prefer to be more scientific and use the word OVUM, which rhymes with "ho-hum." When a Roman soldier called for his breakfast, he might have called for an ovum, because that is the Latin word for egg. Scientists today often use Latin and also Greek words so that they can understand each other,

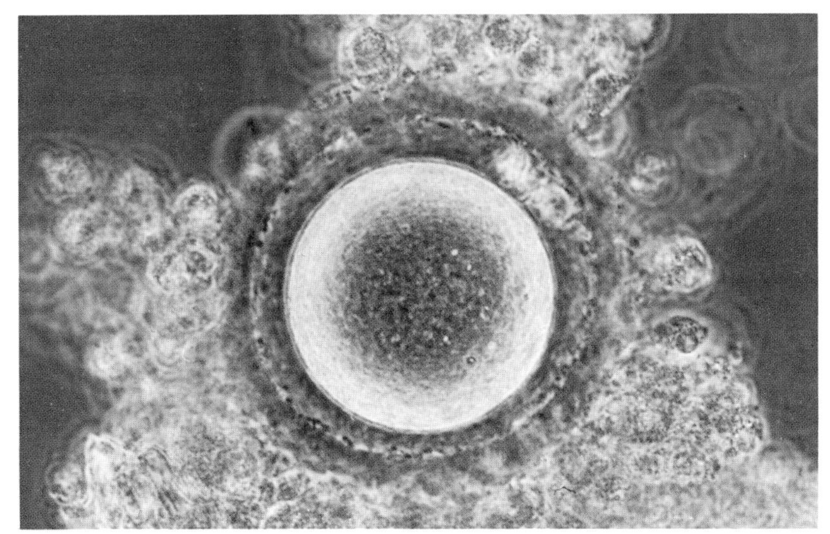

Human egg, magnified about 300 times. The inside of the egg (whitish ball with darker center) is visible through transparent shell; it contains living part with food. Egg is drifting in cloudy fluids.

although they may come from different countries and speak different languages. Therefore, in any language, a scientist would say that you grew from an ovum, a human ovum, and the chick grows from a chick ovum, and a tree grows from a tree ovum.

Large, small, or smallest, the marvelous thing about all eggs is that they don't get their babies scrambled. The right one comes out of the right ovum, or egg, with its head on straight, and with every other part made as it should be, according to its parents. Each ovum follows what we might call instructions that are given in that tiny living dot from the parents.

In this book we shall look into a hen's egg, because that egg is easy to see and easy to find. We shall be able to see how the dot can grow, first looking like a cluster of bubbles, then molding itself piece by piece into the animal it is supposed to be — in this case a chick, in your case you.

DAY
0

The Snowflake Chick

On the yolk of every hen's egg there is a small white patch looking like a snowflake, and inside it is the dot where a chick might begin.

To see the patch, you could open an egg from your kitchen, as you see in this photograph. But wait! First you should know that there is one great difference between your kitchen egg and the egg, or ovum, of this book. No chick could ever grow from your kitchen egg, because the eggs sold for eating usually come from farms where the rooster is kept away from the hens. The

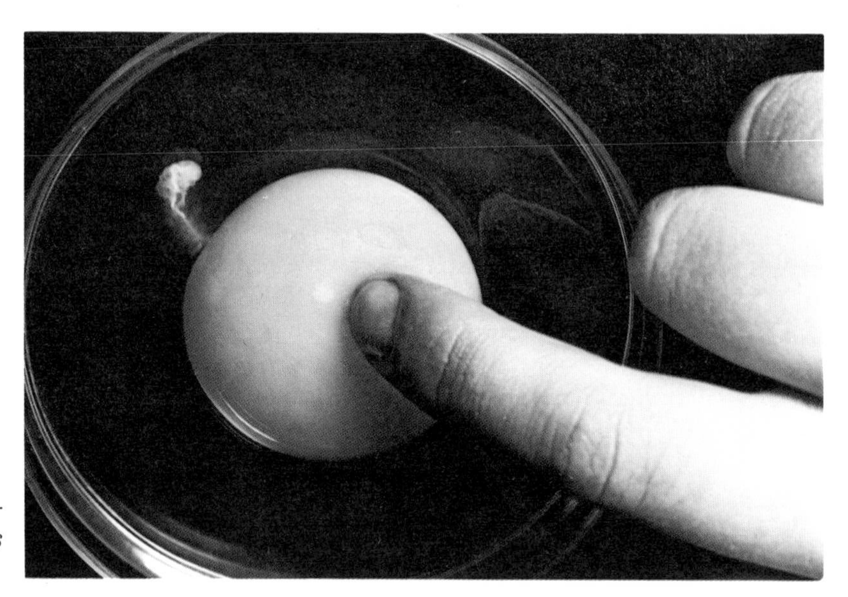

"Snowflake" patch is at fingertip. It floats up if yolk is left undisturbed.

white spot you can see on the yolk of your kitchen egg is merely the place where a chick might have started if this egg had been FERTILIZED. What is fertilized?

For an egg to be fertilized, the mother hen must first of all meet a rooster. This must happen before the egg is laid. It must be before the hard eggshell is made, when only the bare yolk, with its white patch, is lying within the warm protection of the hen's body. Then a rooster might mate with the hen and leave his seed near the yolk.

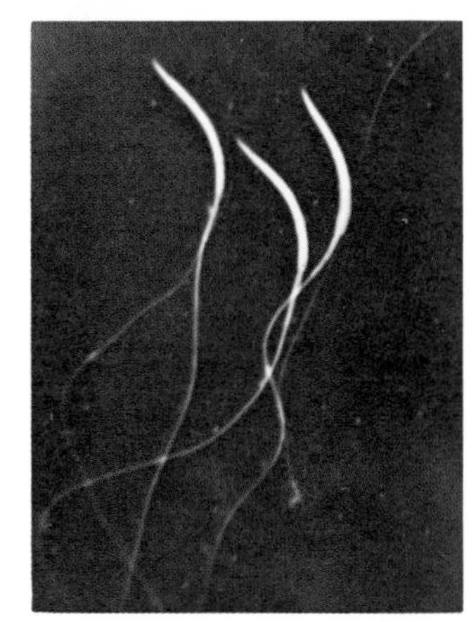

Three rooster sperms, magnified a thousand times.

The rooster's seed, made by his body, is called SPERM. His sperm is much smaller than any egg. It is much, much smaller than this comma, and you could never see the sperm without a powerful microscope. As you can see here, in a photograph taken through a microscope, each rooster sperm has a threadlike tail. Swishing its tail back and forth, the sperm can swim inside the hen and it can swim up to the yolk. Usually many swim up together. One sperm reaches the white patch on the yolk first; that one moves on, right into the white patch, to its center which is called the egg NUCLEUS, pronounced *nóo-klee-us*. Nucleus is the Latin word for kernel or nut.

Egg nucleus (pale, football-shaped oval near center) magnified more than a thousand times.

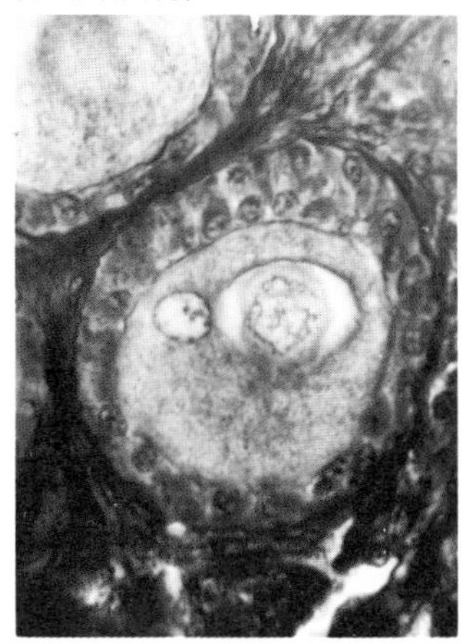

Near the middle of every egg there is such a nucleus, but it is not hard like the kernel of a fruit; it is soft like jelly. The hen's egg nucleus lies concealed and protected inside the white patch on the yolk. The nucleus is so small that you could see it only through a microscope. If you could look at it through a microscope when the rooster's sperm has swum up to it, you would first of all see that the sperm drops its tail on arrival by the side of the egg nucleus. The head of the sperm is itself mainly a nucleus, a mate for the one from the egg. The scene inside the white patch would appear as you can see it in the photograph below. Then the egg nucleus and the sperm nucleus will move closer and closer together, finally to be joined.

When the sperm nucleus joins the egg nucleus, the two become completely mixed with each other, like two raindrops running together. One sperm plus one egg nucleus equals one single new parent nucleus. One plus one equals one: this special addition of sperm to egg is what we call fertilization. It changes the one mother egg to one mother-plus-father egg, and at the same time it seems to wake up the egg to start building a chick.

All eggs, the biggest and the smallest, including the ovum that made you, are fertilized in a similar way. The ovum of a plant, too, is usually fertilized through a male (or "father") plant particle, called pollen, which is fine as dust and is often brought to the plant's flower by the wind or on the fur of a honeybee.

The newly "aroused," fertilized egg nucleus is the beginning for a new life. But where does the new animal come in? There is nothing like the form of a chick in the chick nucleus. How can a chick grow from it? Scientists ask the same question, but

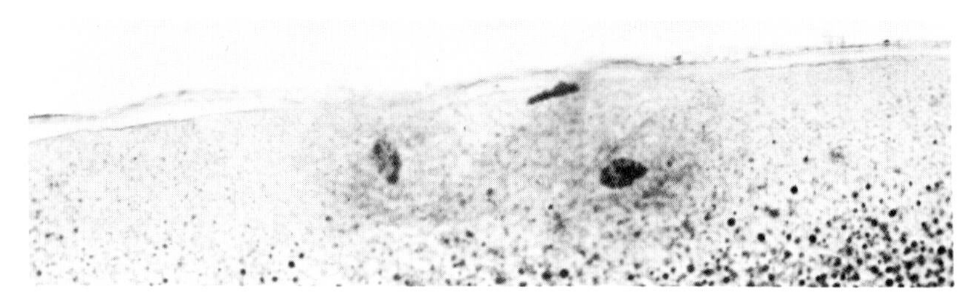

Sperm nucleus (left) approaches football-shaped egg nucleus (right) inside white patch, here greatly magnified.

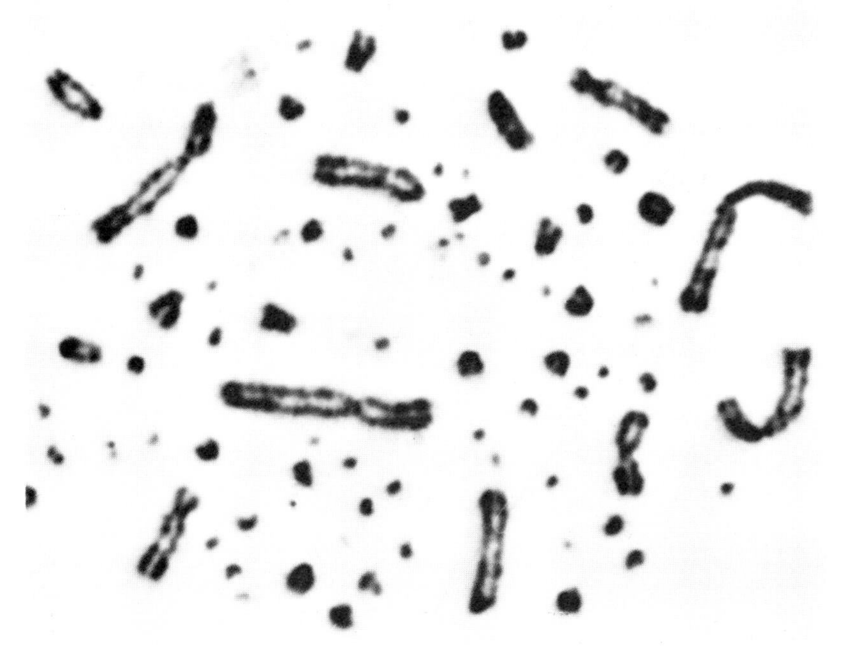

Chromosomes from a chick nucleus, magnified three thousand times. The strands have a wide size range; each pair has one strand from sperm and one from egg.

they have not yet discovered the whole answer. It will have to be found inside the nucleus.

Inside the nucleus there are instructions for making a chick. The instructions are carried by microscopically small strands called CHROMOSOMES, pronounced *kromo-soams*, which comes from two Greek words that mean "color bodies." The chromosomes got this name because they take on a color stain from a special dye that makes them visible under the microscope. But the smallest of the chick chromosomes are hard to see. The total number is not yet known with certainty; it is thought to be thirty-nine chromosomes before the egg is fertilized. Each sperm from the rooster also carries the same number of chromosomes. At fertilization, each strand from the egg pairs up with its particular partner strand from the sperm. Altogether this makes seventy-eight chromosomes for the new chick, one half from the mother and one half from the father. It is a most important fact that fertilization in any egg always brings together a nearly equal number of chromosomes from father and mother.

In coming together, the egg and sperm chromosome strands not only pair up, they also cross over each other, and trade parts with each other. Sections of the mother's chromosomes are traded for sections of the father's. A great deal of this sort of exchange goes on until the composition of all the chromosomes is rearranged. They are then no longer like the chromosomes of either father or mother. They have their own entirely new composition.

Seeing the trading of the chromosomes is only the beginning of the story. The strands that we can see under the microscope are like a face seen in the dark, you can only see the outlines but not the features. What still remains invisible, even through our best microscopes, are the most remarkable portions of the chromosomes. Although they have never yet been seen, some things are known about them. They are called GENES from the Greek word that means "to be born." The genes determine what is to be born. They direct that this will be a chick, not a monkey. They direct that it shall have a chicken brain and chicken heart, two eyes, two legs and one beak, big wings or little wings, yellow or brown feathers. Every single thing needed will grow according to the directions of the genes, thousands of genes along each chromosome. And these thousands of genes are the combined ones from the hen and the rooster. They are combined when the egg is fertilized.

The instructions from the genes are often called a code, like a coded message. Each life is formed according to its own particular message, which is composed out of the messages from its parents. It is difficult to learn how the message works, because it is so different from anything we might write or read or hear. One might say that the instructions are as different from a chick, as musical notes written on a page are different from the sound and shape of the instrument that will play the music. And you know that the notes to be played by a grand piano are no larger in the music book than the notes to be played by a small piccolo. This may help you to understand that the coded message for a bear need be no larger than the instructions for a chick.

The instructions in the egg's nucleus have to be very exact. For our chick, in addition to instructions for making the whole body, inside and out, including every feather, there are also instructions on what the chick will do, how it will hatch from the egg, how it will walk and eat and grow up. Perhaps, as you read along and see how a chick is made, you might have some ideas of your own of how these remarkable instructions might work.

But before we look into our egg to see how a chick is made, let us see how the egg itself is made and then is laid by the hen. As you remember, when the ovum is fertilized, it has no shell around it. Only the yolk lies there, deep inside the hen's body,

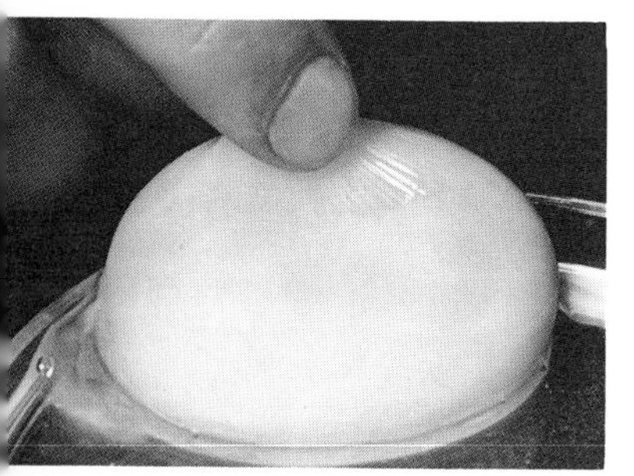

Finger touches yolk envelope.

in a special egg tube. The yolk is enclosed by a very fine transparent envelope that holds the ball of yolk together and also covers the snowflake patch on it. The envelope is so thin that the rooster's sperm can get through it and can reach the white patch, which is called the BLASTODISC, in Greek meaning "growing disc." The yolk and the white patch and the envelope around them, these three are the true egg, or ovum. The rooster's sperm must reach this ovum inside the hen while the ovum is still bare, before the egg white and the eggshell have covered it.

Egg white and eggshell are extra food and housing for the chick. They are made on the day after the ovum has had its last chance to be fertilized. They are made while the ovum, fertilized or not, slowly travels outward inside the soft tube from which the hen will lay the egg. The hen's tube is like a factory assembly line. As the egg glides through the upper part of the tube, a thick, gooey blanket is wound around it. This is the egg white. As you know, it is colorless and gooey when raw, and becomes more solid and white when cooked. The scientific name for it is ALBUMEN, which means "white" in Latin.

Some parts of the albumen become thicker than others. The thickest parts appear whitish even before cooking, and you can easily find them in any raw egg. They look like two small, white, twisted ropes. They are attached to opposite sides of the egg yolk and they help to hold the yolk in place, like ropes for a swinging cradle. The ropes have a strange name; they are called CHALAZAE, pronounced *ka-lát-se*, a Latin word meaning "hailstones." They are called this because the man who named them thought they looked like hailstones.

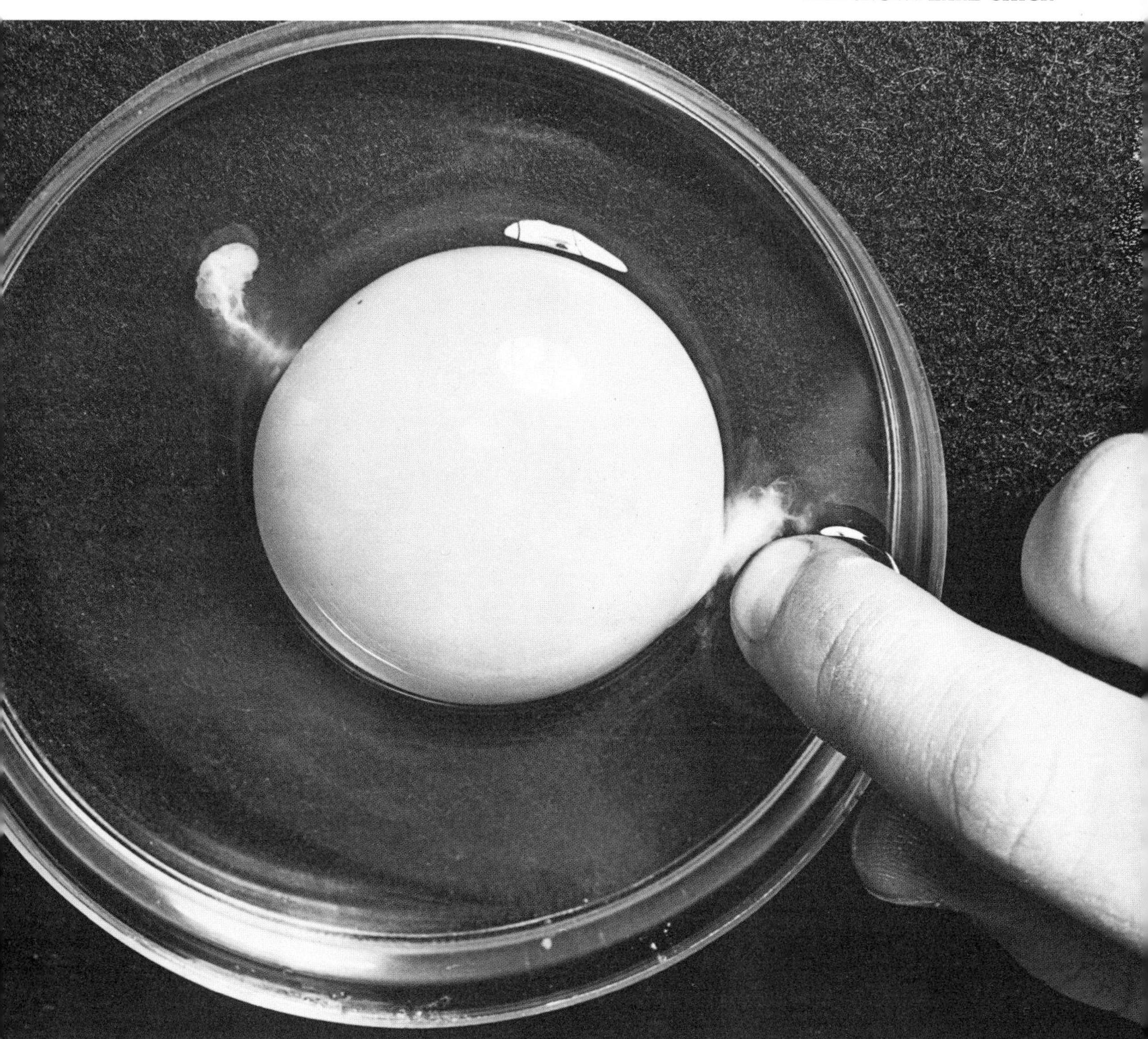

Two ropelets of egg white, called chalazae, are attached to yolk's envelope.

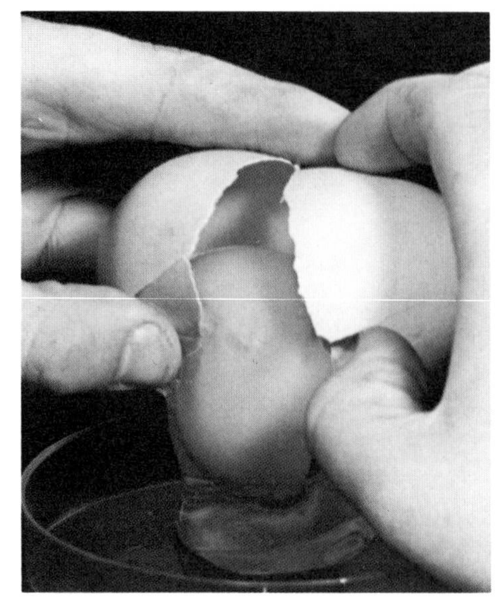
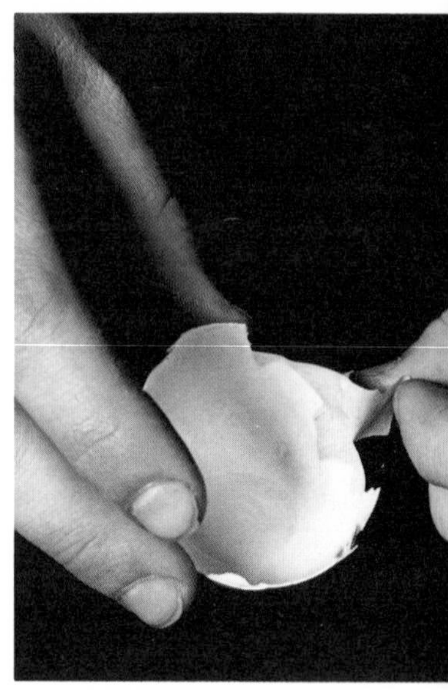

Breaking the egg so that yolk slides out gently, keeps yolk intact. Adhering to shell (at right) are soft shell membranes; if dried, these become papery.

When a good thick padding of egg white covers the yolk, two white envelopes form around it. These are called the SHELL MEMBRANES. They are like a skin. You can easily see these inside any eggshell, although the two shell membranes stick so closely together that they appear to be just one envelope.

At the very end, three layers of shell are made to spread over the egg and to become hardened. Then the egg is ready to be laid. The three layers of shell are pressed so closely together that they appear as a single shell. Although the shell will be hard, air and moisture will pass through it after it is laid. Perhaps you have noticed when you tint Easter eggs that the inside of the egg may also become colored. This is because the watery tints pass through the eggshell. In the same way, air and moisture pass through and will be available to the growing chick.

A special bubble of air comes into every egg as soon as it is laid. If you would like to see this bubble, break a kitchen egg into a cup and look inside the broad half of the eggshell. You will find a bulge formed under the whitish shell membrane. Poke it gently with your finger. It is a bubble of air that has come in through the eggshell and has been trapped under the shell membrane. Tear the bulging membrane with your fingernail, and the bubble will collapse because the air escapes through the hole you made.

Every hen's egg has this special space to store air, called the AIR SPACE. As you will see at the end of this book, that air space will become very important to our chick just before it hatches.

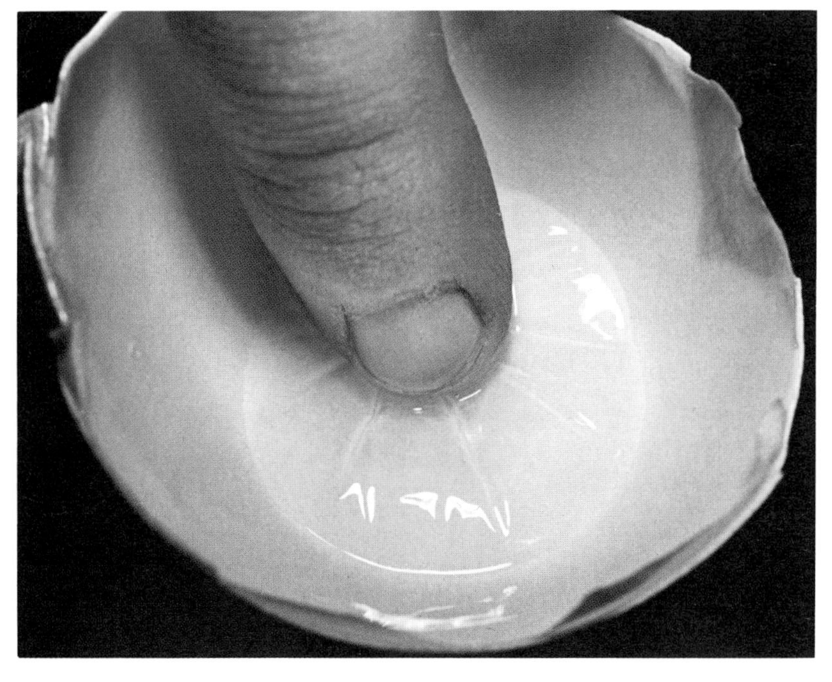

Finger pokes air space.

DAY
1

The Chick Is Made of Cells

Our chick begins its life aboard the egg that is slowly traveling outward within the hen. If you could watch this egg through a microscope, you would see exactly what these photographs show you, because they were taken of just such an egg.

In the first photograph you can see a split near the center of the snowflake patch, which as you already know, is called the blastodisc. That split is, one might say, the blastodisc's blast-off for our chick's life. It means that the one fertilized egg nucleus has divided in two. At that moment it turns into two nuclei (nuclei is the plural of nucleus) that are the first two nuclei of the new chick. In about half an hour a second split appears, and it forms an "x" with the first one. This means that there are now four nuclei. Each of the four sections of the "x" contains one nucleus, but the nucleus is much too small to be seen in these photographs. In seven hours, about as long a time as your school

First split: short line at center of spotty "snowflake," magnified about 20 times.

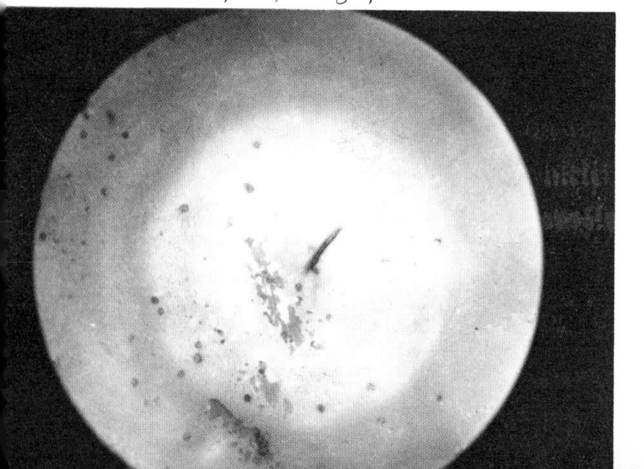

Second split: "X" creates four divisions.

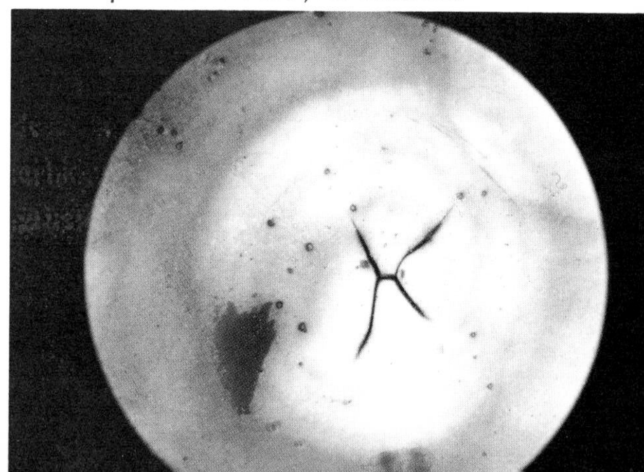

day, three dozen divisions and in them three dozen separate nuclei have formed. When the three dozen divide they make six dozen and so on; the number then increases quickly.

By the end of the first day, you might be able to count more than 200 divisions. But, as the number grows, the size of each division gets smaller. How can something become smaller as it grows? Think of a cake: if you cut it into four pieces, you get rather large slices. But if you would go on to cut it into 200 portions, your slices would become very small. In the same way, the portions of the blastodisc become smaller and smaller. Now they look like a mass of tiny bubbles lying on the yolk. Unlike cake, these bubbles are delicate and fragile, and they are living. If you could observe them for a while, you would see movement among them as new ones are formed, as they shift positions, as some rise to the top and others move toward the middle. Yet, if you would move away from the microscope, you could not see these divisions or this activity. Without a microscope, your eyes could see only the plain white patch, very slightly larger than the whitish spot on your ordinary kitchen egg yolk.

In the past, before people had microscopes, they did not know

Thirty-two divisions: after seven hours.

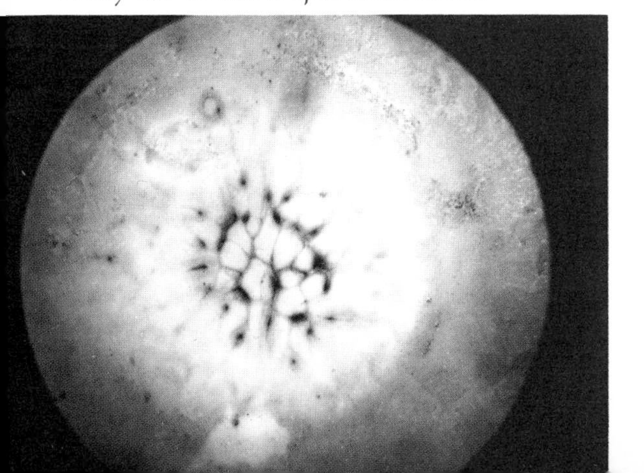

Near end of Day 1: 256 divisions.

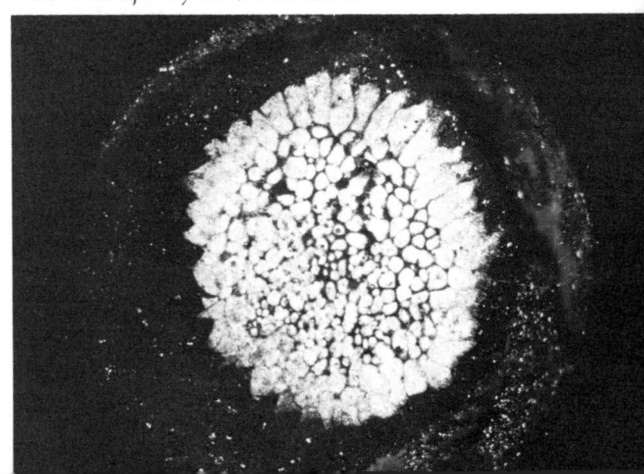

that these bubbles existed. Today we know that chicks, and all living things, are made piece by piece out of these bubble-like, living building blocks. They are called CELLS. The word "cell" comes from the Latin word *cella* which means "little room." Each cell, in addition to being a kind of building block, is in a way like a little room in which the instructions from the parent cell are carried out. Each cell has a nucleus, and each nucleus contains the full set of instructions. Every time one cell divides in two, the full instructions are copied for the new cells.

Some very simple forms of life are made of only a single cell. Perhaps you have heard of an amoeba, which is such a single-cell animal that lives in water and can be seen only through a microscope. To make a new amoeba, the old one divides in half. Each half becomes a new amoeba and goes its own way.

All the more complicated living things are made of great numbers of cells that work and live together. You are made of about two hundred million cells, many times more than the number of stars you could count on a clear night. If you could look at any part of yourself under a microscope, you would see cells, different-looking cells depending on the part you were seeing. Since you began from one cell, you too were once two cells, then four cells and so on. A week after your life began, you were probably about two hundred cells. You grew more slowly than the chick and, of course, your cells were different from the chick cells. They were special human cells with special instructions to make *you*.

Our chick, when it is one day old, already has several hundred cells. At this time it is about to land in the hen's nest, surrounded by nourishing yolk, well wrapped in egg white, well protected in its newly made eggshell, and usually having its arrival well announced by the loud cackling of the mother hen.

Although the chick in the egg is so well provided and protected, it still needs some mothering after the egg is laid. The chick cannot grow if the egg is abandoned. For the next three weeks it will need a mother hen to sit on the egg, to keep it warm, and to turn it over occasionally by nudging it in the nest. If the egg is taken away from the nest, it must be kept equally warm in an incubator and must also be turned over at least once or twice a day to be able to grow without its mother.

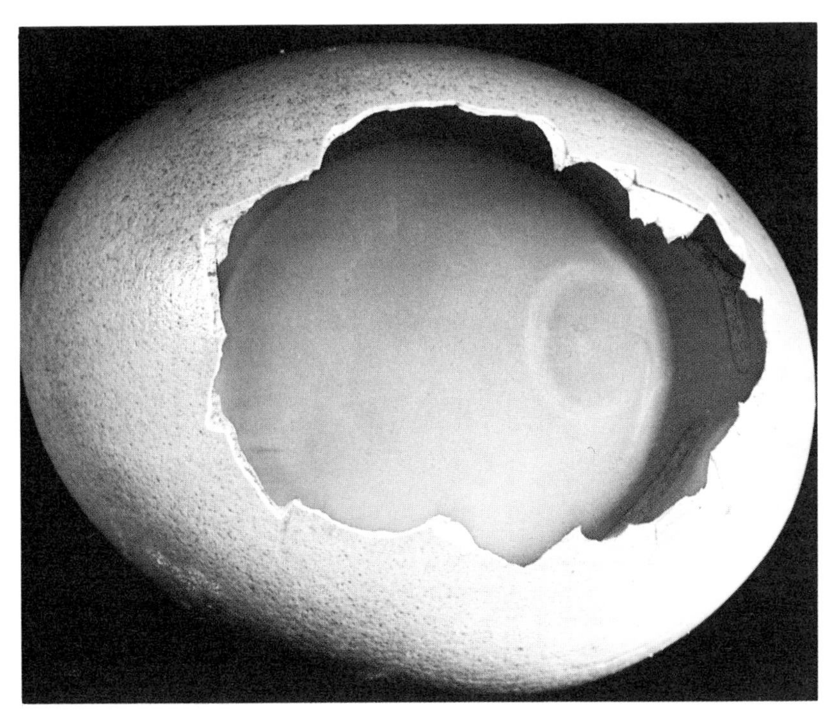

"Snowflake" disappears under a raised halo a few hours after egg is laid and kept warm. This is caused by water drawn into yolk envelope from egg white.

DAY
2

A Body, a Beating Heart, and a New Name

On the second day of its life, our chick is transformed. Yesterday it was a mound of bubble-like cells that you could count. Today thousands of new cells grow and arrange themselves into the strange form you can see in the photograph at right.

Does it, perhaps, remind you of folded petals of a flower bud? Or does it look to you more like a shoe? Actually, what you see here is the very beginning of a body. What may look like the top of a shoe will become the head and heart of the chick, the front of the shoe shape will grow into the rest of the body, and what appears like buttons will become backbone and muscles.

This certainly does not look at all like a chick. It has a special name. It is called an EMBRYO. It is no longer a fertilized egg cell, and is not yet a chick—it is a chick embryo. Embryo is a Greek word, and it means "grow within." The embryo is usually so fragile that it must grow within, either within the warm protection of the mother, as you did, or within an eggshell as a bird does, or in the case of plants, within a seed. Each chick, and every life that begins from one cell, must first grow into an embryo as its body is made bit by bit, and each part is made to fit and work together with every other part.

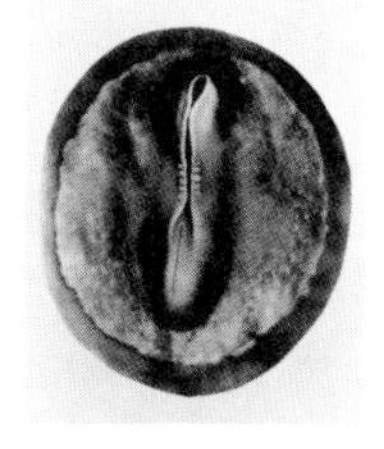

Real size.

If you could sit and watch this chick embryo through a micro-

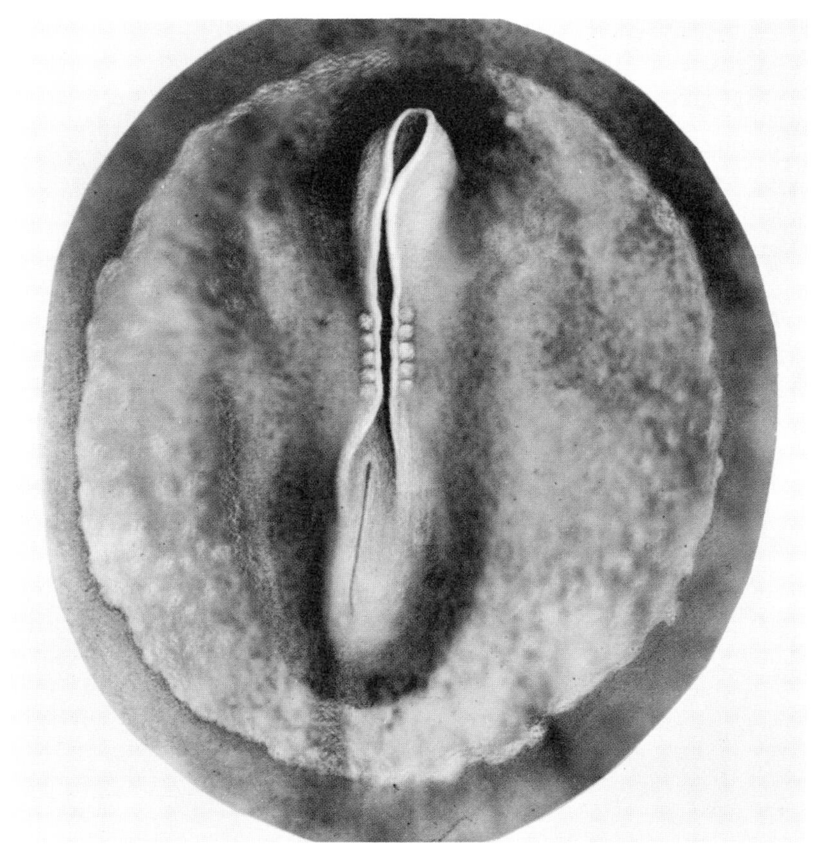

Day 2: Magnified. Upper folds will form head, brain, heart; "buttons" will be muscles and backbone; lower folds grow into rest of body.

scope for a few hours, you would be amazed to see how alive it is. To see it well, you would have to look at it under great magnification. Yet, even when the embryo is so enlarged, each of its cells appears no bigger than a pinpoint.

Watching through the microscope, you would see a continuous slow-moving traffic of these pinpoint cells. You would see streams of cells, some moving forward, others toward the center, others turning under. In the same places where there was only

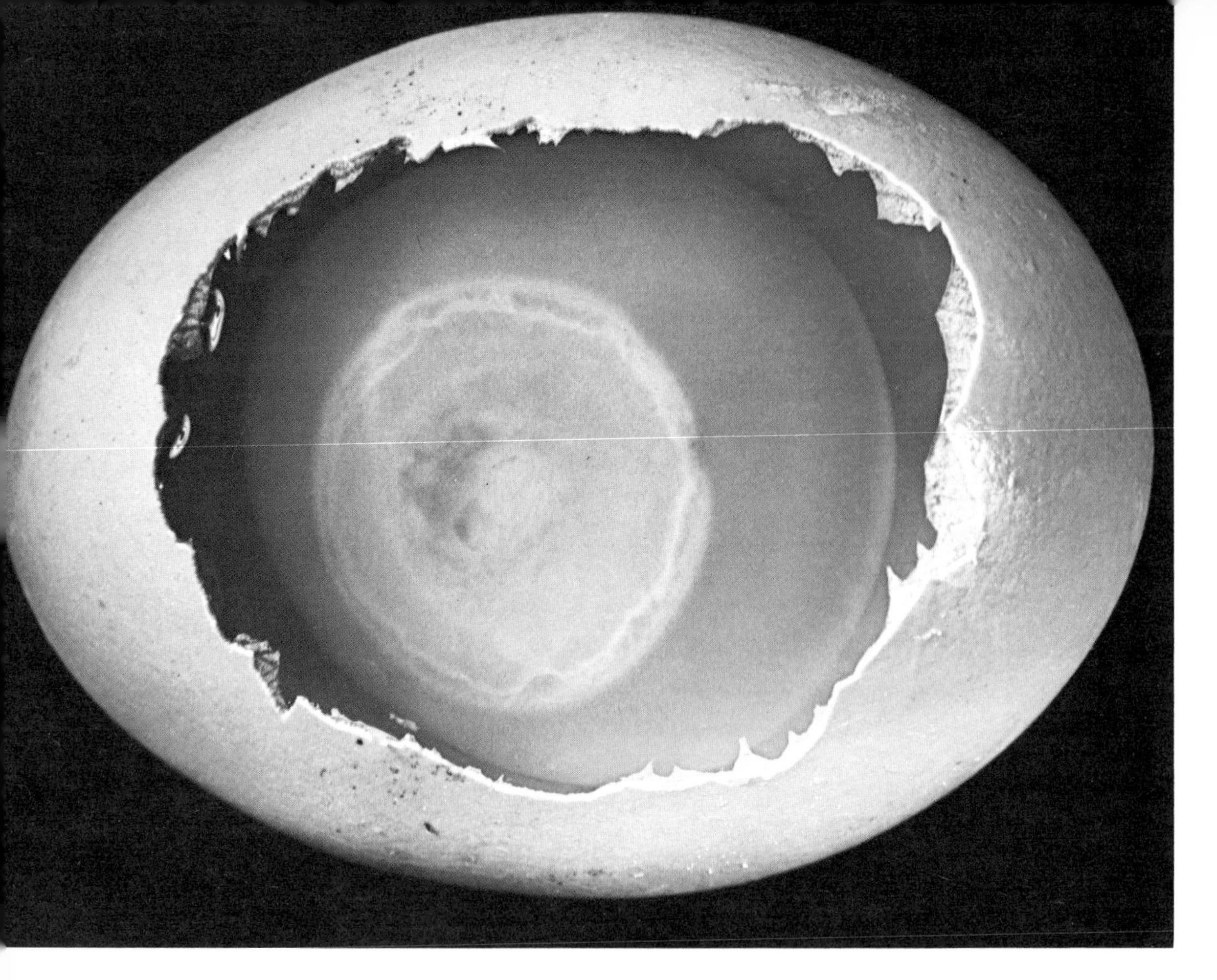

A footprint shape, at center of spreading cloudiness on yolk, indicates that an embryo is growing here. "Footprint" is the outline of a light area around embryo (magnified opposite).

plain yolk two days ago, a great activity has now been stirred up. The activity is busy but not wild. Each cell goes exactly to its proper place, probably according to the instructions it carries in its nucleus, and also because it is guided by the other cells around it. The cells travel in groups, and eventually a group may settle in one place. When the group arrives at its destination, it stops traveling and stays to build the special part for which it is intended.

The chick embryo, like all embryos, grows according to a strict plan. About six hours after the fertilized egg has been laid, if it has been kept well warmed, you will find that the chick-making cells in the egg have arranged themselves into the form of a shield, called the embryonic shield. What the cells will do next is so much the same in every hen's egg that scientists have been able to predict exactly what each group of cells will go on to build.

If you continue to watch the shield under the microscope, you will next see an extra number of cells lining up along the center of the shield to form a thick band, looking like a streak, as you can see in the photograph on the left, below. This is called the PRIMITIVE STREAK, which means beginning, or earliest, streak. More and more cells gather around this streak, especially toward its upper end. So many cells stream forward that the head of the streak becomes a crowd of cells. These cells are packed closely together, and this makes the head of the streak look darker than the rest. Gradually the darkened crowd of cells fans out in an arrow shape. This is the very beginning of the brain. Then, as the arrow shape grows out farther, it will be the beginning of the heart.

Embryo, magnified about fourteen times, begins with primitive streak (left); grows into arrow shape (center) with first "button" forming at its middle; about three hours later there are four pairs of "buttons" (right).

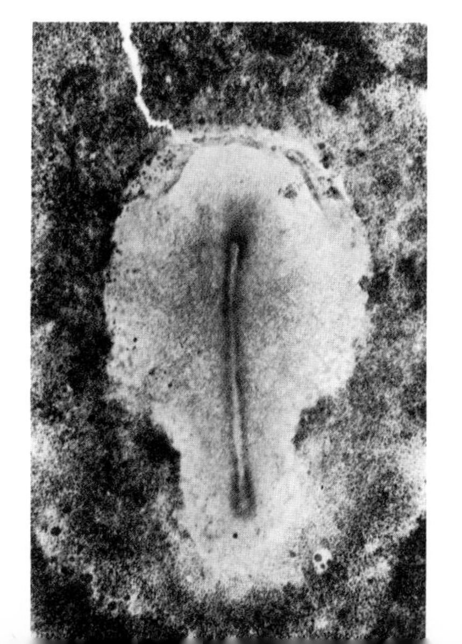

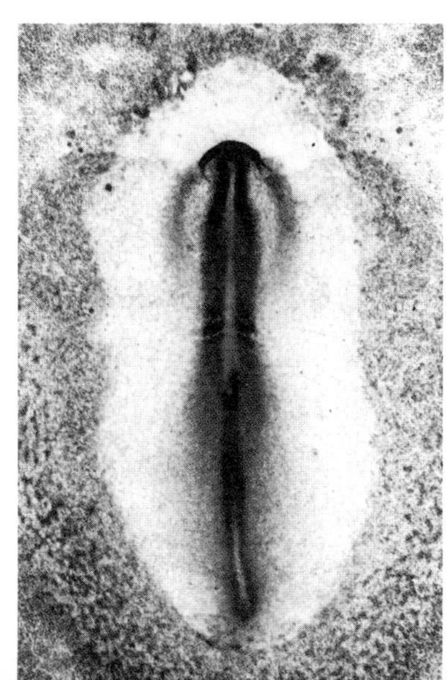

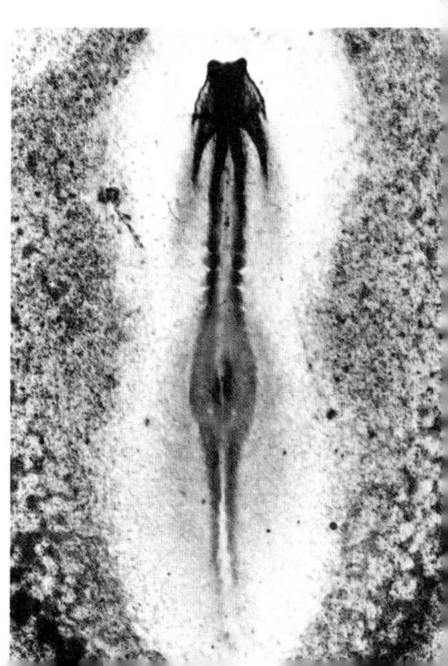

To see all this in the photographs on these pages, you have to look at them closely and carefully. Understanding these photographs is somewhat like learning to read a new kind of printing. Only people who are used to looking at embryos through a microscope can understand right away what they are seeing. Scientists who study embryos know, from experience and from experiments, that the part of the streak that looks darkest will go on to form the brain, the head, and the heart. The part of the streak that looks lighter will become a hollow tube, and in it groups of cells will gather to form the other inner parts, such as the stomach, the intestines, the liver and kidneys.

At the same time certain cells wander away from the main groups of the embryo. They move outward onto the yolk. These cells will create blood for the chick. Blood, as you may know, is a liquid filled with many cells. If you look at red blood under a microscope, you can see that it is actually a colorless liquid, filled with a variety of cells. Some of them have no color, others have a rusty red color. These are called red blood cells, and they are the ones that make the blood appear red.

To make blood, the chick's wandering cells establish tiny cell islands on the yolk. The islands become surrounded by small amounts of colorless liquid. Then, in a few hours, some of the island cells form a substance called HEMOGLOBIN, a Greek word that means "globes of blood." This hemoglobin is mostly made of iron substances from the yolk. The color of hemoglobin is rusty red. This gives red blood cells their color, and that is how red blood appears out of the yellow yolk.

Opposite: Day-2 embryo is mainly head and chest. This is enlargement of photograph on page 31 (lower right); it is the same stage of development shown on page 29, under different lighting. Four pairs of somite "buttons" are recognizable in all three photographs.

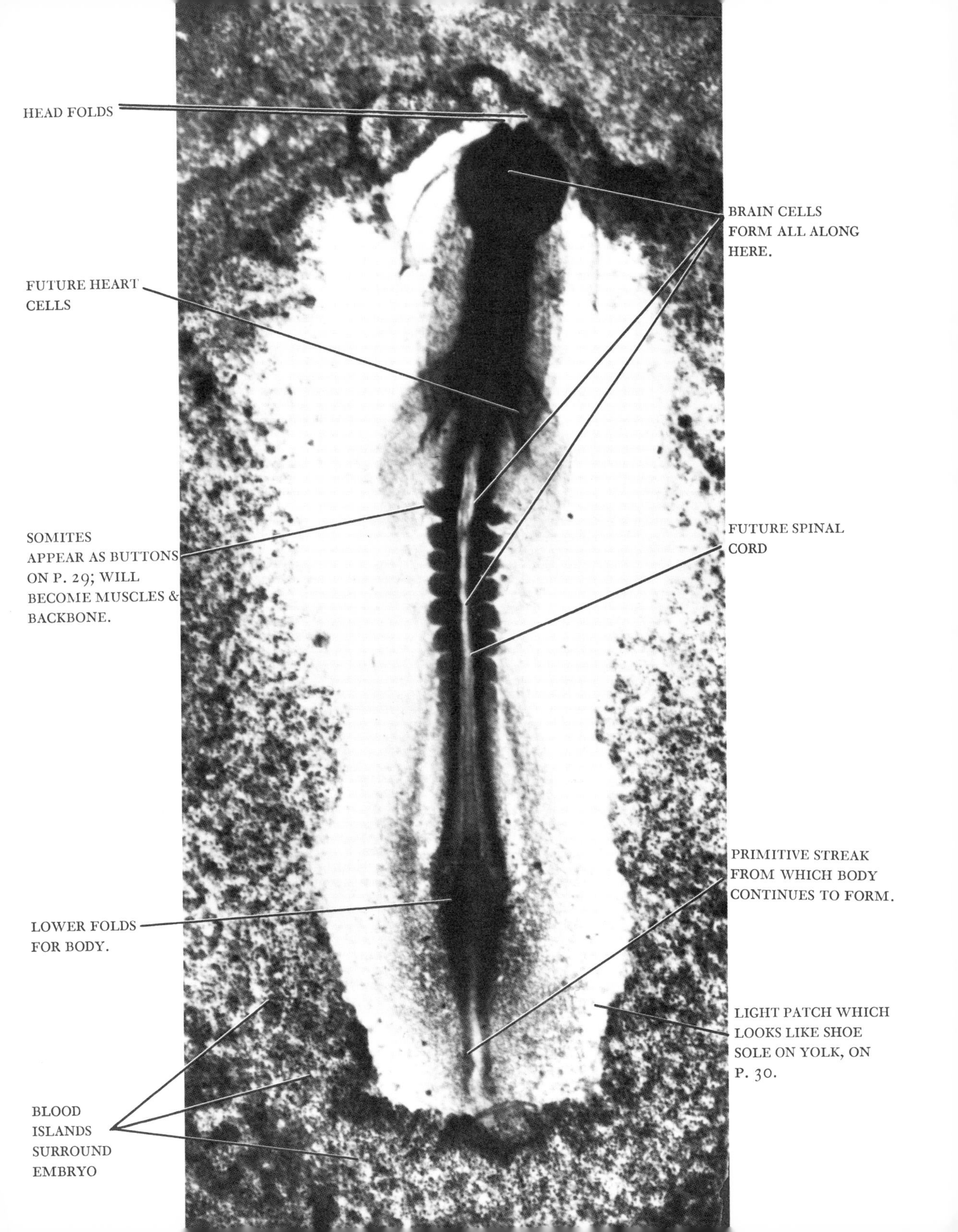
HEAD FOLDS
BRAIN CELLS FORM ALL ALONG HERE.
FUTURE HEART CELLS
SOMITES APPEAR AS BUTTONS ON P. 29; WILL BECOME MUSCLES & BACKBONE.
FUTURE SPINAL CORD
PRIMITIVE STREAK FROM WHICH BODY CONTINUES TO FORM.
LOWER FOLDS FOR BODY.
LIGHT PATCH WHICH LOOKS LIKE SHOE SOLE ON YOLK, ON P. 30.
BLOOD ISLANDS SURROUND EMBRYO

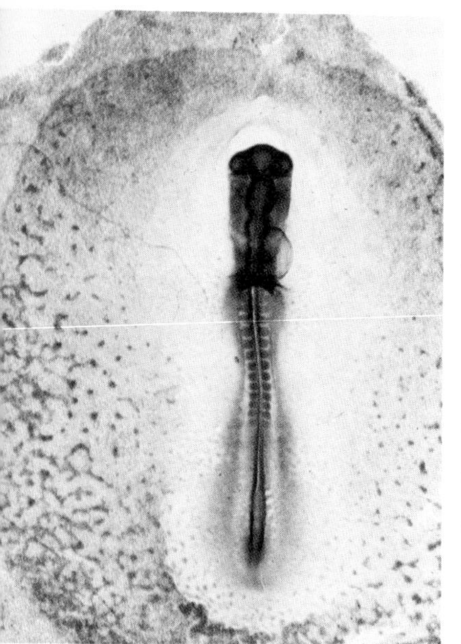

Head folds have joined at top. Heart has begun to beat; it looks like a pouch (upper right). Somite "buttons" have increased to thirteen pairs. Outer dark spots are blood islands.

While the blood is being made, other cells build soft walls around the island pools of blood, keeping the blood from flowing out. Later these walls grow into soft tubes called BLOOD VESSELS. As the blood vessels grow, they branch out over the yolk, like branches of a tree, and are connected to other blood vessels that have meanwhile grown within the chick embryo.

Inside these blood vessels, the blood will carry nourishment to the chick. Tiny particles of the nourishing things yolk is made of, such as sugars, fats, vitamins, proteins, iron and other minerals, can pass into the blood vessels. They pass through the walls of the vessels in a way that is similar to roots of a plant taking in nourishment from the soil.

The nourishing bloodstream will flow along to the chick because it will be pumped by the chick's heart. While the blood is forming, the heart is becoming ready to pump. At first it merely twitches every now and then. After working like this for two or three hours, the heart starts to pump quite regularly. Then the moment comes for the largest main blood vessels to open up into the heart, and for the bloodstream to begin to shuttle in and out of the heart.

Through a microscope it is an impressive sight to see this first opening up of the bloodstream traffic to the heart. At the opening moment the large vessels, looking like super highways, begin to carry a stream of little dots, which are the blood cells, into the heart. Two other large "highway" vessels carry the cell traffic from the heart to be distributed throughout the embryo.

Blood flows into the heart when the heart muscles relax, and is forced out when the heart muscles tighten up. This is how the heart begins to pump; and it will never stop for the whole life of that chick. When you look into the egg tomorrow, you will plainly be able to see that beating heart.

Expanding circle of cloudiness is still only plainly visible sign that embryo is growing here.

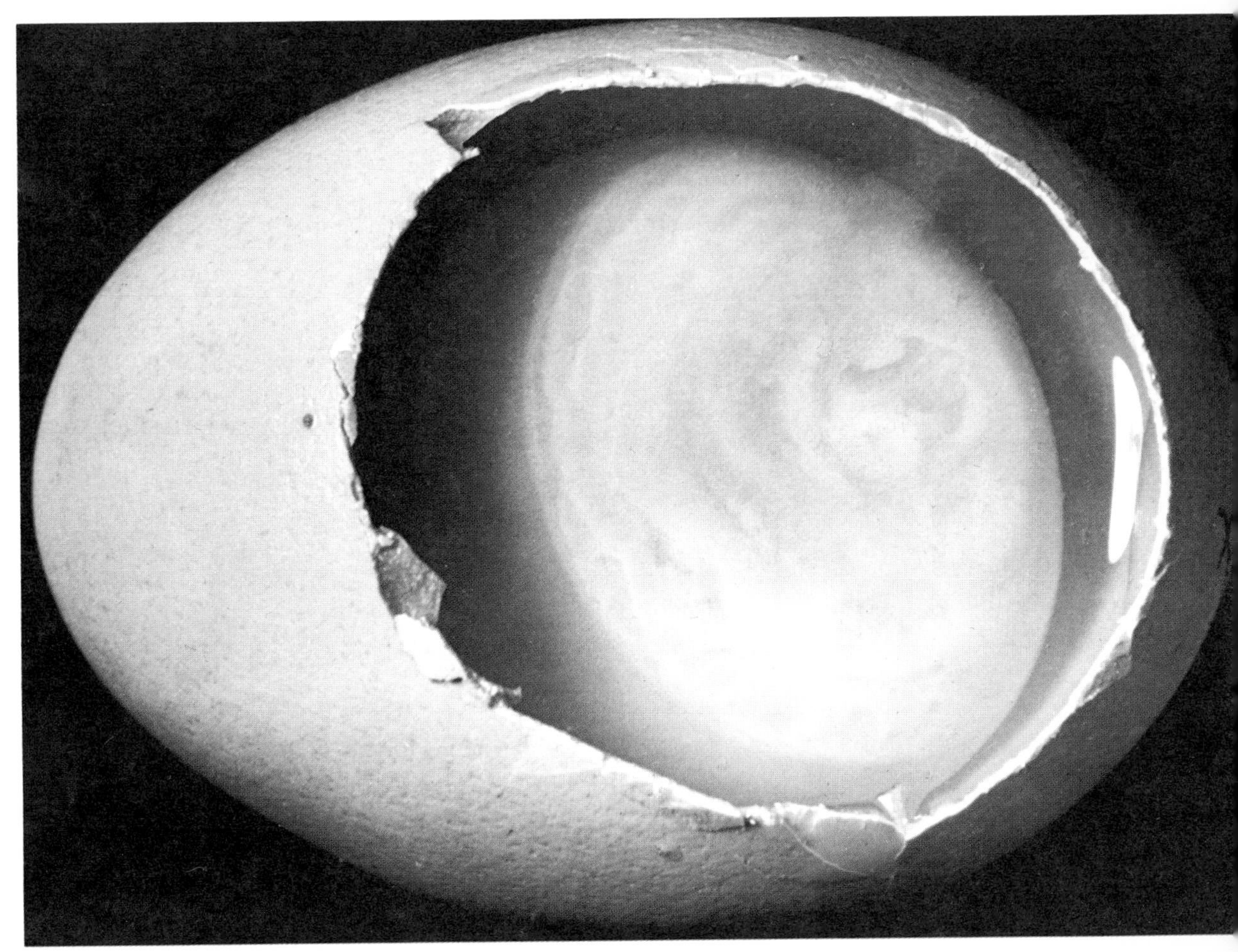

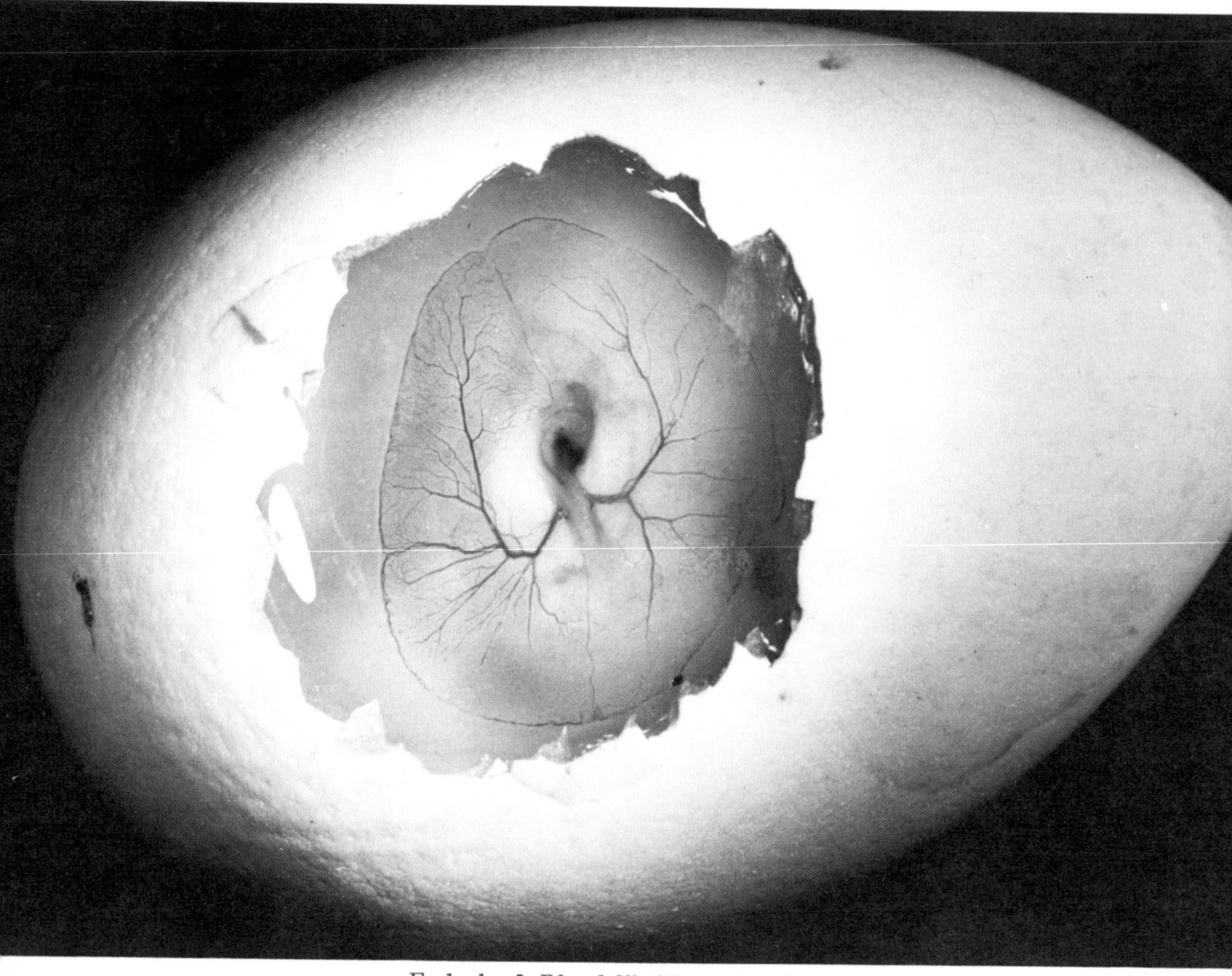

Early-day 3: Blood-filled heart, beating more than a hundred thirty times a minute, is heart-shaped at this instant. Embryo outline resembles top half of S, or question-mark in reverse. Winglike branches are blood vessels on yolk.

DAY 3

The Question-mark Chick

The marvelous surprise in the egg today is a tiny throbbing chick's heart, ruby red on the yellow yolk. Like the pumping movement of a bellows, the heart beats vigorously and regularly, telling you—here is life.

For the first time today you can also plainly see the bright red branching blood vessels that have grown up around the embryo. As you already know, the heart pumps blood, bringing nourishment from the yolk to the embryo through these vessels.

The body of the embryo which surrounds the heart is still so delicate and so transparent that you can only barely see its faint outlines, as you can try to see them in these photographs. At right the embryo is close to the size it would be in real life.

The outlines of the embryo have changed a great deal since yesterday, but this embryo still does not look like a chick. However, while it does not yet look like a chick, it does look very much like most other kinds of embryos, including all birds and all mammals. All these animal embryos, even the embryos of people, look very much alike at the beginning. This likeness among embryos shows us that all animals begin from very similar, simple foundations for a body. Because the chick embryo happens to grow in an egg that is large and easy to get, it is one of the favorite embryos for scientific study. In finding out how the chick grows,

Mid-day 3: One and a half times real size. Heart is at center.

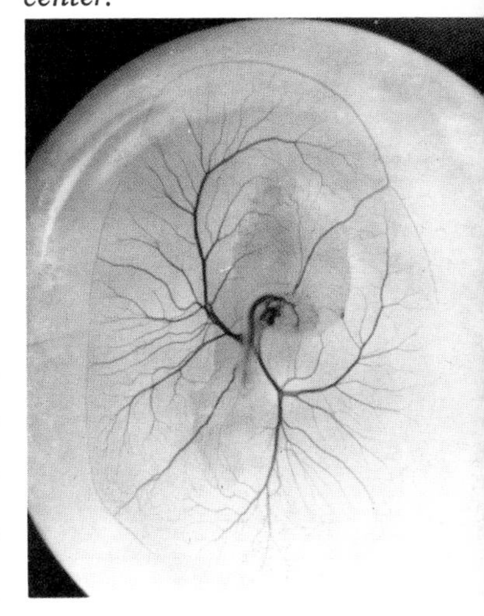

one can find out a great deal about all of life, because all living things have much in common.

The making of a living body is never like making a machine such as an automobile, where all the parts are first made, then assembled, and finally made to go. An embryo is not put together, it forms itself. All embryos have a life of their own from the start, and the more important parts for being alive must form before the less important ones. The shape of the embryo reveals to us the order of importance of its different parts. The brain has a head start in growing, therefore all embryos have rather large heads. The heart begins immediately after the brain, and it soon forms a large bulge right under the large head. A hard-working heart is needed very early to pump the embryo's blood, carrying nourishment to the surrounding community of cells, so that these cells can continue to increase in number to complete the rest of the body.

Unopened egg in front of light, revealing shadow of embryo and blood vessels. Air space is at top.

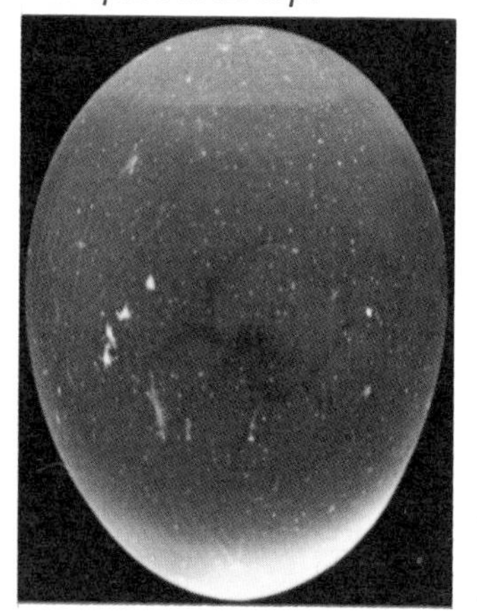

For a while, the embryo is mainly head and beating heart. The rest of the body is merely a slim, transparent tube. Within it, the other inner-body parts begin to grow, including the stomach, the intestines, the kidneys, the liver, and the breathing parts. These inner parts of the body are called the VITAL ORGANS, from a Latin word meaning "life" organs, or parts, because they are so important for being alive. It is only after these all-important vital parts have begun to form that the time comes for the rest, including muscles, bones, skin, and limbs.

Some embryos, such as fish and tadpoles, come out into the water at this stage before the whole body is finished. They can swim, they have gills and can breathe and live on their own under-

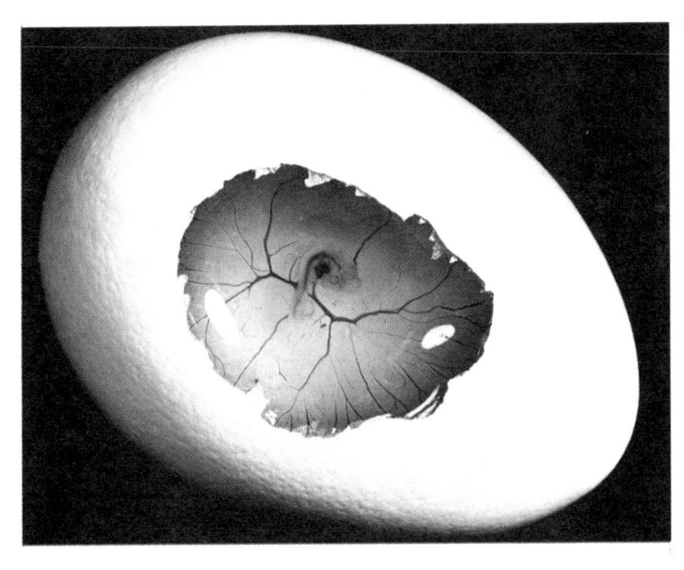

Late-day 3: Eye coloring first appears in tiny grayish disc, to lower right of heart. Embryo lies on side with large head bent far downward. A few hours earlier (previous page) eye was not yet visible.

water. That is the important point; they are at home in the water, and water gives them a certain kind of protection.

When you learn to swim, and find that you can float, you discover that water gives support. For this reason fish do not need to grow such strong bones as land animals. Even more important to an embryo is another kind of protection that water gives. You can understand this best if you think of a delicate petal of a flower. If you put such a petal into a cup of water, it will float gently, and will stay fresh and moist for several days. If you try to poke the floating petal with your finger, it will drift away unharmed. But if you lay the petal by itself on a table, or even on a soft cushion, you can bruise it easily with your finger. In any case, the petal left out in the air will soon dry up and crumble.

All embryos are so delicate that they cannot exist without the protection of water. Before they can live out in the open, surrounded by air, touching and bumping into things, they must have a quite complete body with strong bones and a good skin. A human baby's skin is often called its birthday suit. Until we are ready for this day of birth, we must grow within, you within the mother, the chick within the egg. And there we all grow within our own small pool of special waters.

On the third day in the egg, the chick embryo begins to show preparations for this pool of waters. Through a microscope you can see a colorless and transparent sheath like a plastic raincoat growing to cover the body. You can see it in the photograph at right. Underneath this sheath, clear waters will accumulate, drawn from the surrounding egg yolk and egg white, so that the sheath will soon be like a water-filled balloon. The sheath is called AMNION, a Greek word that means "little lamb." It got this name because little lambs are often born with a part of this sheath wrapped over their heads. The lamb and you and the chick, and millions of others, begin life inside an amnion, gently bobbing about.

The water-filled amnion is made of cells. Along with the cells that build the chick embryo, many other cells grow to build the special structures needed inside the egg to take care of the embryo. As you have seen, there were cells that wandered out onto the yolk to create blood and blood vessels, and now there are cells to make the amnion. The amnion and the vessels on the yolk will be left behind in the egg when the chick hatches. Although they do not become a part of the chick, all these extra things had to be included in the original instructions in the first cell of the chick.

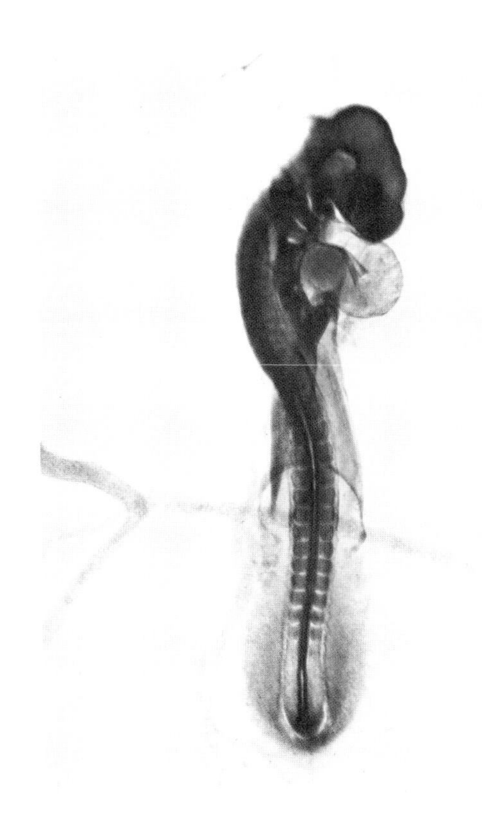

"Question-mark" chick, magnified about eight times, is same age and shape as page 36. Cupped place for eye is dark; heart is partly transparent pouch on chest; amnion is short skirt; below it are more somite "buttons."

According to the instructions, the chick embryo lies flat, face down on the yolk, until it is two and a half days old. Then, slowly, it begins to turn over to lie on its left side, showing us in profile its narrow body, its bulging heart, and one of its two eyes on the forward-bent head. Lying on its side like that, the embryo seems to form a question mark in reverse. All embryos turn to this reverse question-mark shape. Could they be asking, "What am I?" From this day on, each kind of embryo will begin to reveal what it is. The different ones will no longer look alike. Our question-mark chick, in one more day, will be unmistakably a bird.

DAY 4 DAY 5

A Bird Appears in the Egg

On the fourth and fifth days a very small figure with large eyes greets us from the egg. The large eyes say, "I am a bird." They answer yesterday's question, "What am I?" Only the embryos of birds have such fast-growing eyes. Therefore we can now be certain that this embryo belongs to the bird family.

It is not yet obvious that this will be a chick. It could be any bird. But it no longer looks like the embryo of any totally different animal family. Along with the particularly big eyes, its body is also becoming birdlike. It begins to have a stub of a bird tail, and it has the beginnings of legs and wings. At this stage, each kind of embryo begins to show its special features that tell us what it will be. You, for example, had a particularly fast-growing brain, and your whole head was your most prominent part for a while.

When the chick has grown this far, it begins to move. It rocks back and forth quite regularly and appears to be nodding its head. The chick is not yet able to rock itself; it is being rocked by a motion of the amnion bubble that surrounds it. There are some threads of muscle among the cells of the amnion. The muscle threads tighten up and then again relax, moving in and out in a back-and-forth motion which sets the chick to rocking. It is thought, although we do not know entirely why, that the

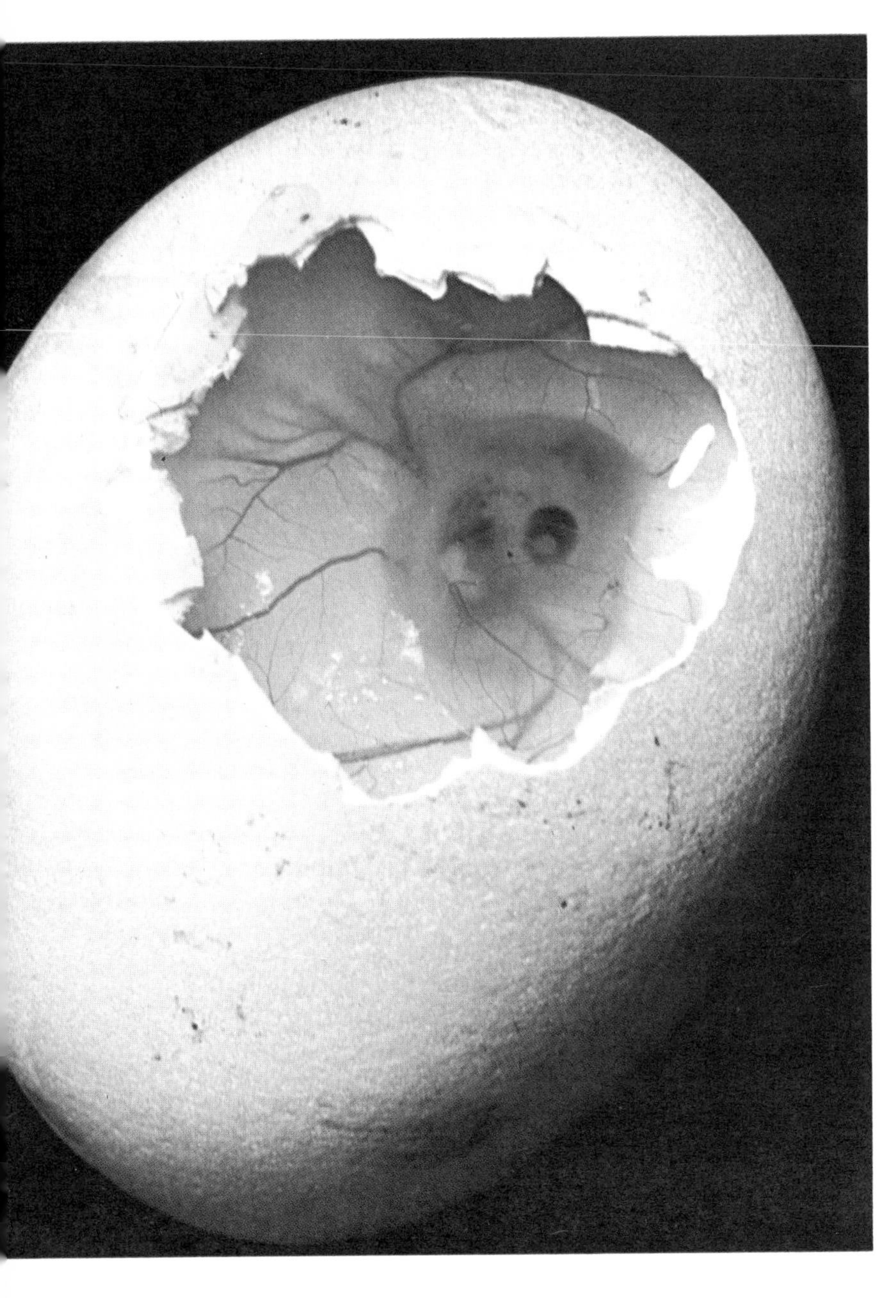

Chick lies on its side. Beneath round, dark eye is the darkish blur of the beating heart. In middle is pale, stubby beginning of wing.

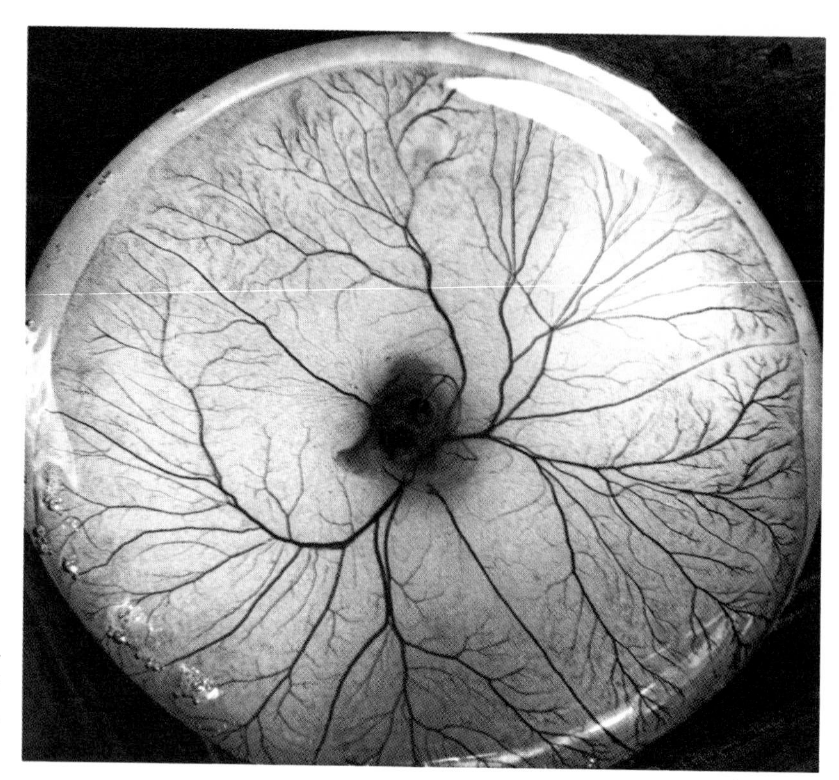

Day 5: Real size. Blood vessels run through transparent yolk sac. Spread-out yolk appears large.

rocking helps the chick to grow properly. Later on, the movement will become much greater, as you shall discover.

Meanwhile, the chick needs more and more nourishment as it grows larger. Looking at the chick here, one might think that it could just reach out for a nibble of yolk whenever it needed some. But, as you know, the chick is completely sealed off from the rest of the egg by the amnion bubble. Furthermore, the bubble is filled with clear fluid so that the chick lives underwater. Although the yolk is so nearby, it requires quite a complicated process to get nourishment to the chick.

This is how the feeding process works: You can see that the chick is lying in a web of blood vessels that has grown over the yolk. Do you remember that these blood vessels are like roots? Particles of nourishment are taken into them and are carried in the bloodstream to the embryo.

As the chick grows, it develops a special tube on its belly through which it receives the bloodstream from the yolk. All the blood vessels from the yolk join and lead into this tube on the chick's belly. But you must not imagine that the bloodstream sloshes around inside the chick. It always runs inside blood vessels. The blood vessels from the yolk lead directly into blood vessels inside the chick's body. The embryo has its own blood vessels that branch out within the body and reach every part in the chick.

The tube on the chick's belly is similar to such a tube that babies have before they are born. On babies it is called the umbilical cord. On chicks it is called the YOLK STALK, and you can see it on page 46. When your umbilical cord dropped off at birth, it left a scar, which is your navel. The chick's yolk stalk will also leave a small scar that will be the chick's navel, and will always be a reminder that this chick was once in an egg.

Together with the stalk, a YOLK SAC also grows. This is a transparent sac, like a clear skin, that grows over the ball of yolk. This sac contains the blood vessels on the yolk. Many new ones are constantly growing in keeping up with the growing needs of the chick. Most of the chick's food comes from the yolk, but some also comes from the egg white, or albumen, through a similar system

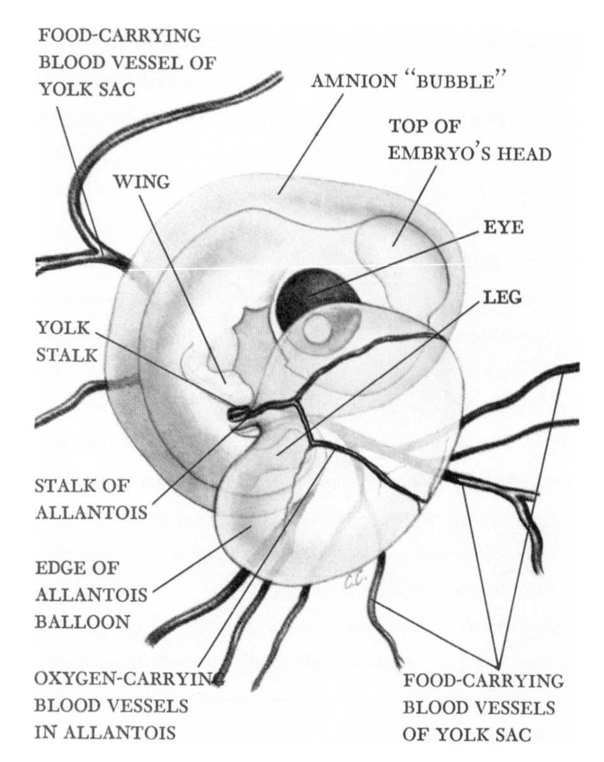

Drawing shows chick contained within amnion and connected, through yolk stalk, to blood vessels on yolk. A separate connection leads to allantois.

of a sac and vessels. While the yolk is mainly made of foodstuff with a little water, the albumen is mainly made of water with a little foodstuff. During the chick's first week in the egg, much of the water is drawn from the albumen, which shrinks down to a small, gummy lump that eventually will be used up.

Food is not all that the chick needs. It also needs air. It needs that part of the air called oxygen. The embryo needs oxygen before hatching as much as the chick will need air to breathe later. Air, and with it oxygen, comes into the egg through the eggshell. But the embryo cannot breathe it, living underwater in its sealed

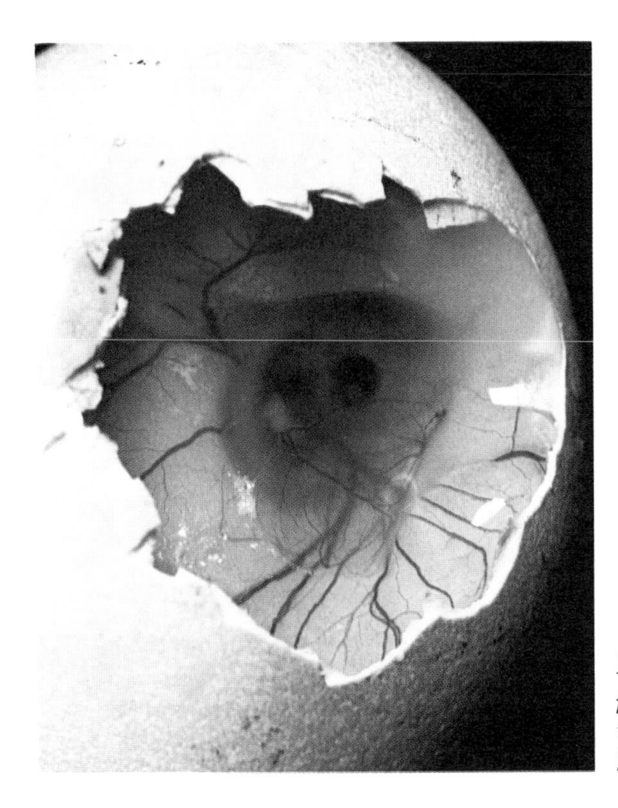

Photograph matches drawing opposite. Eye, wing, outline of amnion and of allantois are visible here, but yolk stalk is not.

amnion. The embryo must receive its oxygen, just like its food, delivered in the bloodstream.

Oxygen comes to the chick through another special system that grows in the egg. One part of this looks like a collapsed balloon, and the rest of it is made up of blood vessels. The deflated-looking balloon is called the ALLANTOIS, pronounced *allan-toys*. Allantois is from a Greek word that means "sausage-shaped," probably because the allantois looks somewhat like the clear skin around a sausage.

The allantois lies near the inside of the eggshell, and it has

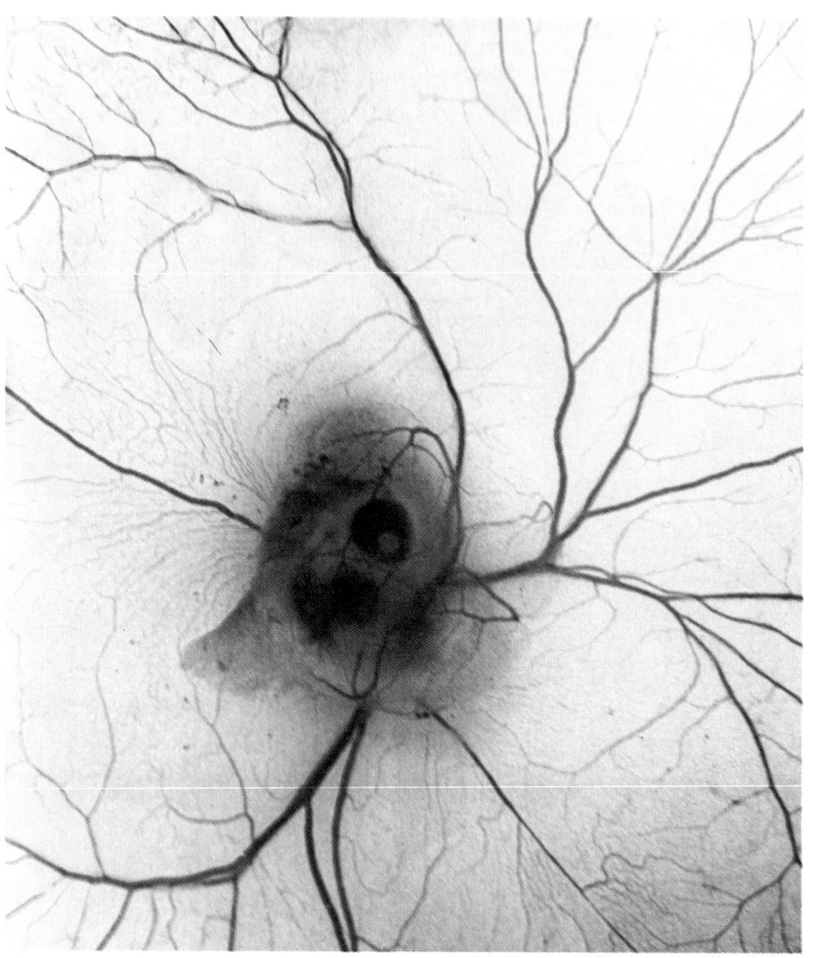

Enlargement of embryo shows small blood vessels. These branch into many still smaller, visible only through microscope.

many blood vessels running through it. The blood inside the vessels of course includes red blood cells, the ones that have hemoglobin, which is the blood's oxygen carrier. Since the allantois lies so near the eggshell, the air coming in through the shell immediately reaches the allantois blood vessels. Oxygen passes into the vessels and is carried to the chick by the hemoglobin in the red blood cells. The shipment of oxygen reaches the chick through a tube on its belly. This tube from the allantois is located right next to the yolk stalk.

However, the chick needs still more, more than food and oxygen. It needs to get rid of that part of the air which we breathe out, which is a gas called carbon dioxide. And it needs to get rid of some of the same substances that will later be in the chicken's droppings. These are the leftover, unusable particles of nourishment. They are waste substances. All the wastes, the carbon dioxide gas, and the tiny particles of food wastes, are removed from the chick through special outflowing blood vessels, in a return stream back to the allantois. From there the carbon dioxide gas escapes outward through the eggshell, but the food wastes stay in the allantois. So the allantois has a double function; in the first place it helps to supply oxygen to the chick, then it also acts as a sort of waste bucket which will be conveniently left behind in the egg when the chick leaves.

Before the time comes when the chick can leave the egg, it has a lot of growing to do. On the fourth day it weighs no more than two postage stamps. On the fifth day it weighs ten times more than that, but even then it is still lighter than one quill feather of a hen. The embryo is so lightweight because it is small and does not yet have any solid bones. You can easily see why it still needs the special protection within the egg. Each day from now on it will become much better prepared to come out.

DAY
7

Enormous Eyes and Another New Name

After one week in the egg, our chick, lying on its pillow of yolk, looks very curious. It has such enormous eyes, even larger than before. This chick has grown a great deal in the last few days. It has become more chubby, padded out with muscles, covered with skin, and it has a better tail, just right for a chicken. The legs have grown long and have four toes, but not yet any claws. And the wings, too, have grown but are folded across the chest where you cannot see them very well.

Only the head of the chick is still very unfinished-looking. But the main parts are there, the brain, the eyes, the ears, the nostrils, the mouth; they merely need to develop more fully to complete a proper-looking chick.

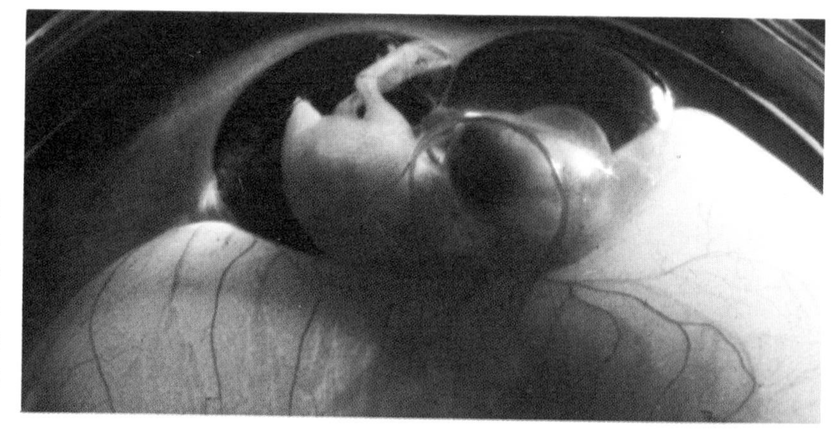

Chick is in fluid-filled amnion bulging out over its body. Blood vessel over chick's head leads from chick's belly to yolk. Head is partly transparent; eyes dark; tail and legs well formed.

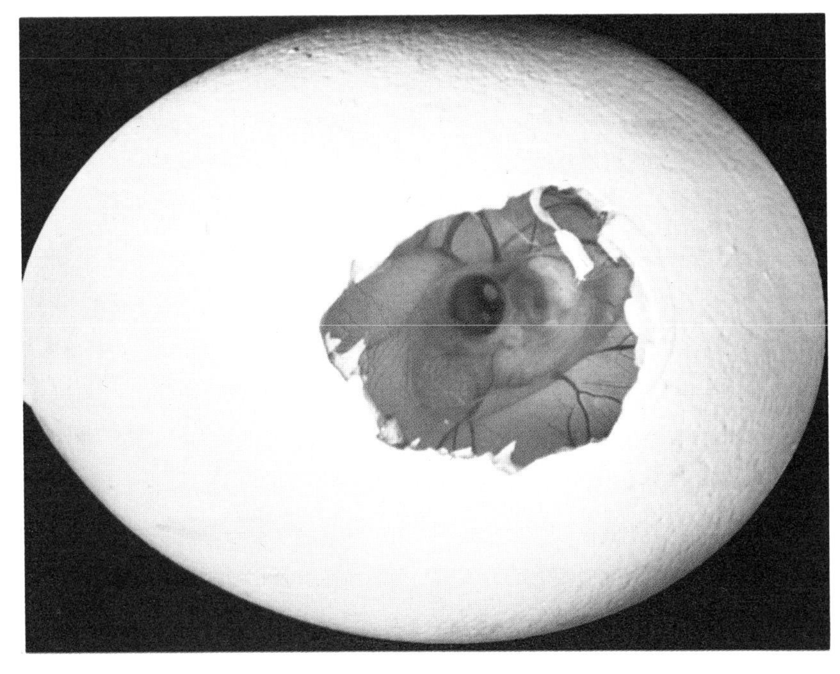

Eyes are always open. Eyelids are not yet developed; eyes are unfinished.

This one-week-old chick is no longer an embryo. From now on, until it hatches, it is usually called a FETUS. The word fetus is the Latin word for "young one." All animal embryos are called fetuses after their most important parts are created. They have this name from then on, until they are born. The chick fetus does not lie lazily on its yolk cushion in the egg. It is doing much more than just quietly growing. Two days ago we noted that it began to rock back and forth. As expected, it now begins to move much more. Almost every minute it exercises its new muscles, moving its head and body, its wings, legs, and tail. It does not have to think about this. The exercising probably happens because, as you know, the chick is constantly being slightly pushed about by the in-and-out motion of the amnion balloon that encloses it.

In the photograph on page 50 you can see this fluid-filled amnion balloon very well, and you can also see that the balloon has grown larger together with the chick. Now the balloon moves more than before, and this not only rocks the chick, it sometimes even turns it over, upside down, and around. Since the chick is floating in waters, its rocking is somewhat like that of a bath toy when you stir up the bath water. But unlike a toy, the chick is alive. When it gets pushed around, it moves automatically, as you, for instance, blink your eyes when something approaches your face. So the chick too moves its head, or wings or legs, when these are pushed slightly. And the chick, too, blinks its eyes in the egg. Although, at this age, it cannot yet close its eyes because the eyelids have not grown large enough to cover the eyes.

We know all about the activities of the chick in the egg, because it is possible to carefully remove a piece of the egg's shell, under scientifically clean conditions, while the chick goes on growing inside. This gives us a window into the egg. Many thousands of eggs have been studied in this way to learn about the growth of embryos. Through such windows it has also been possible to watch how chicks behave inside the egg. Each chick acts very much like every other chick. Most of them seem to do about the same things at the same ages.

In addition to the rocking movements, all one-week-old chicks appear to nod their heads several times a minute and occasionally lift the head up a little. The nodding seems to be caused by the pounding of the chick's heart, which now beats quite regularly about 140 times a minute. The pounding heart might cause the head to nod, as a bumpy car ride might make your head bounce

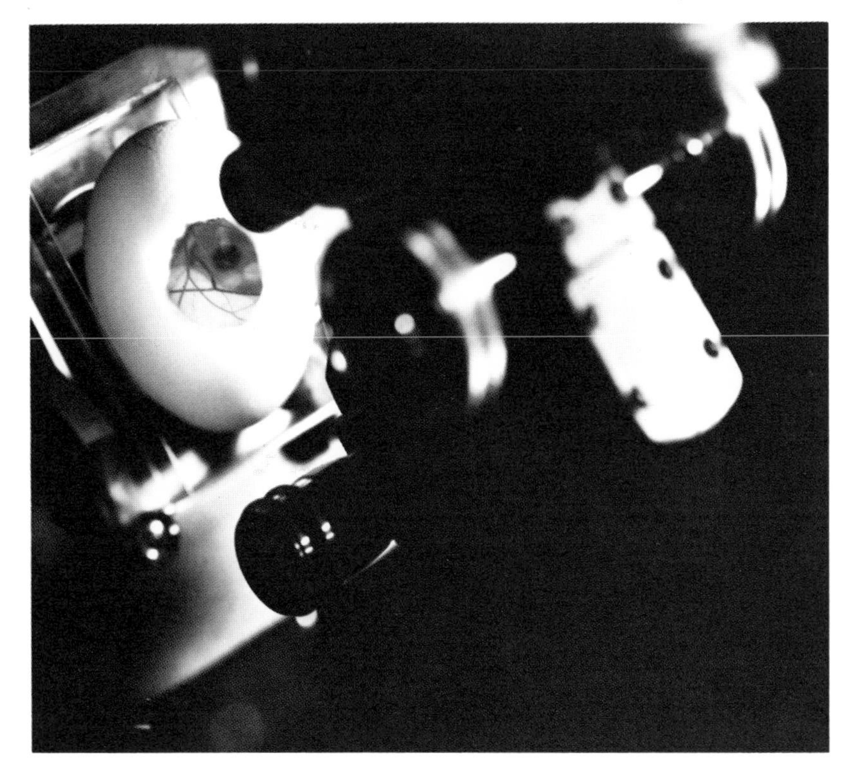

Glass window, sealed into egg, leaves growing chick unharmed. Through microscope one can see fine details such as blood cells inside the blood vessels.

up and down. Then, occasionally, the chick adds a little motion of its own by raising the head and, a day or so later, by turning the head from side to side as often as once or twice a minute. Near the end of the first week, the chick also begins to open and close its mouth, which does not yet have a visible beak on it. The beak only begins to grow out on about the seventh or eighth day. By that time the chick begins to stretch and bend its whole body and legs, and now and then, like a proper bird, it wiggles its stub of a bird tail. Such movements will eventually help the chick to be able to get out of the egg. Everything that happens inside the egg seems to be preparing for that moment.

Chick can move inside fluid-filled amnion, curving overhead, and here reflecting light (top center). To left of tail is yolk stalk; from it two blood vessels lead to yolk underneath chick. Beak is partly hidden by wings. Rows of "goose pimples" on back are beginning feathers.

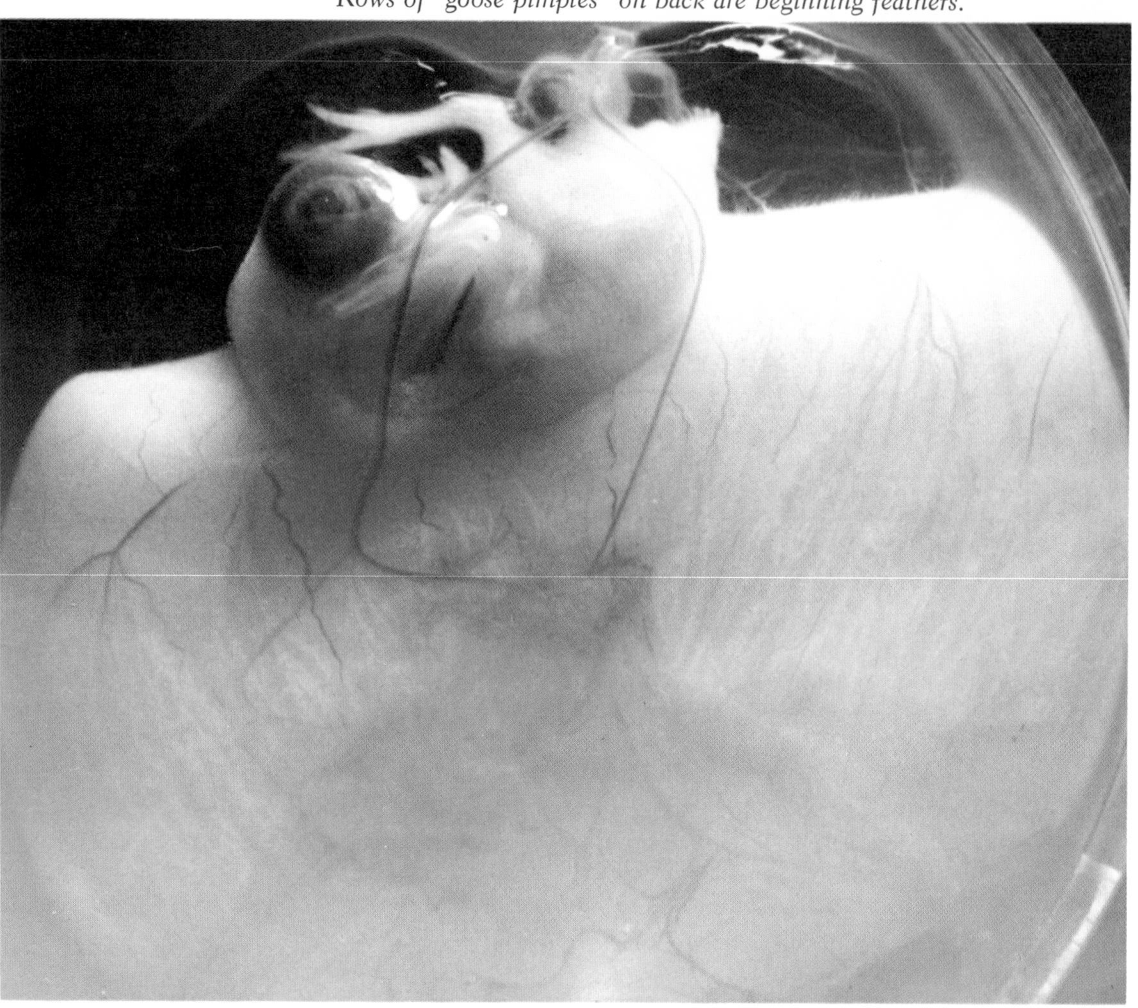

DAY 11

The Chick Gets Goose Pimples

Now the chick has been in the egg eleven days. You have to look closely at this chick to see how it has changed in the last few days. It has not changed as much as it did from one day to the next when it was younger. Its growth from a ball of cells into an embryo was a much greater change, yet that change happened in only a few hours.

The older the chick gets, the slower the changes will be. After the chick has hatched, you may have to wait a whole month to see any big changes. This is true of all living things. The older we get, the slower we grow. When you were a baby you changed every week, now it may take several months to see a difference.

Look again carefully at the eleven-day-old chick. Do you see how its head has changed? It is no longer partly transparent; it is now covered by skin and has the curve of a chicken's head, tapering into a long neck. The eyes are even larger than before, and you can see the dark pupil in the center. Something new has grown on the front of the chick's head, but in this photograph the head is bent so far forward that this new thing is partly hidden by the wings. Yes, it is the beak. It started to grow at the end of the first week. It is still soft, but will harden in the next few days. Sometimes the chick will open and shut its beak rapidly a few times. The chick's legs have grown to be longer, with elegant long

Chick (shadow lower right) is here behind yolk and great blood vessels.

toes and fine claws. Sometimes the chick may accidentally scratch its head with the claws. Then it turns the head away. And every now and then, the chick makes a wriggling movement with the whole body, as if it were not entirely comfortable.

A tangled forest of blood vessels has grown up around the chick to carry the oxygen and food that make the chick grow steadily bigger and heavier. One week ago it weighed less than a hen's feather, now it weighs fifty times more, which is about as much as two pennies weigh. In the meantime the amnion balloon has stopped rocking the chick, so that the chick lies more quietly for a while. For the next week it usually settles into one position, crosswise in the broader half of the egg. This chick is nearly completed. Although its body will become larger and more perfected, and will change in appearance, it has already grown all the essential parts it needs to be a chicken.

But where are the feathers? If you look closely at the chick on page 54, you will see neat rows of something looking like goose pimples on the skin of the chick. These "chick pimples" are the beginnings of feathers. They will grow out into a downy coat that will protect the chick when it leaves the egg. You can already see some short, downy feathers growing out around the tail. There are still nine days left until the time for hatching. Still plenty of time, and yolk, left for the chick to grow bigger, stronger—and smarter.

Two huge dark eyes are at upper left; thin wing is at center. Uppermost blood vessels take in oxygen, below are food-carrying ones. Vessels will dry up and remain in eggshell at hatching.

DAY 14

DAY 15

DAY 16

The Egg Becomes Crowded and

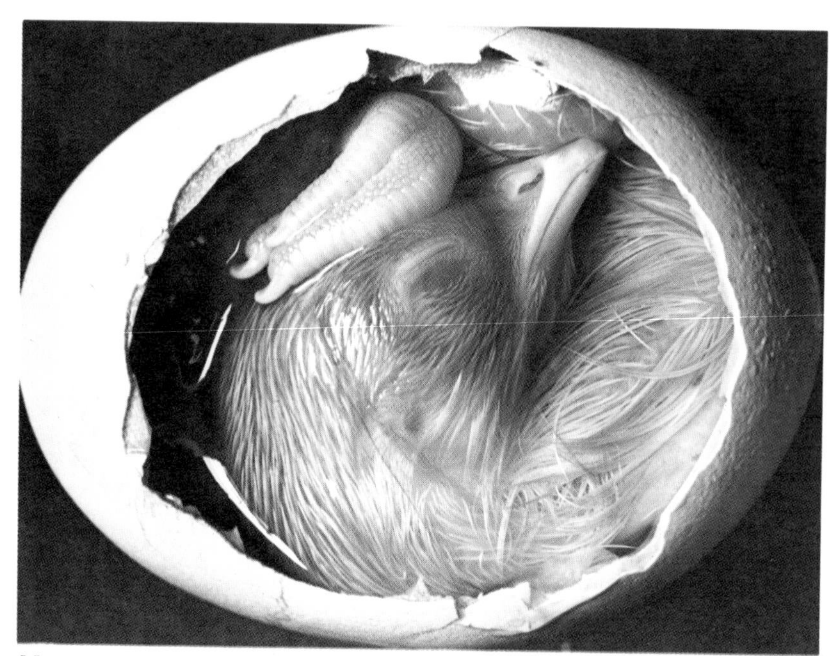

Nearly ready to hatch, chick must lie tightly curled up. Yolk is now used up.

DAY 17 | DAY 18 | DAY 19 | DAY 20

the Chick Prepares to Move Out

In the third week the egg becomes a crowded place, filled with an active, feathered chick. The chick begins to get in its own way. Its wings are tightly folded over its chest, and the long legs are folded up over the wings, with the toes sticking right into the chick's face. As you shall see, the chick can move its toes, scratch its beak and head, and turn its head from side to side.

But the chick is now too cramped to be able to make any large movements. Now and then, as before, the chick gives a wriggle or makes a jerking movement, as if it were getting restless. But it looks sleepy with its eyes closed most of the time. If a light shines on them, they open wide and then close again, as you can see here. At times the chick also opens its beak wide and shows its thin pointed tongue. More often the beak opens only slightly and is quickly shut again, and this is repeated in quick succession. It is a movement like that of your lips when you say "mamamamama." When the chick does this, it is drinking. It swallows small gulps of the surrounding pool of waters. If you have ever watched any

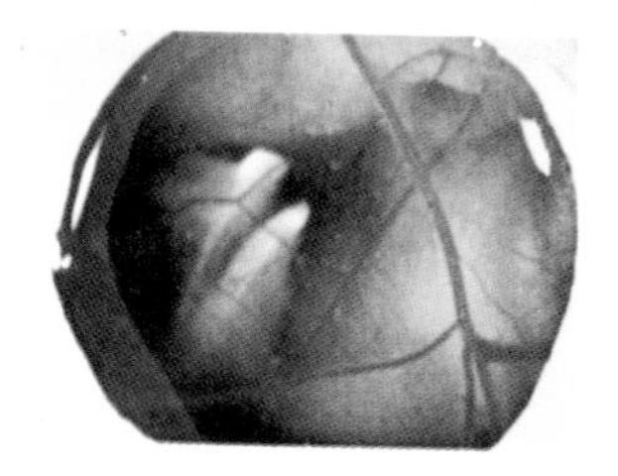

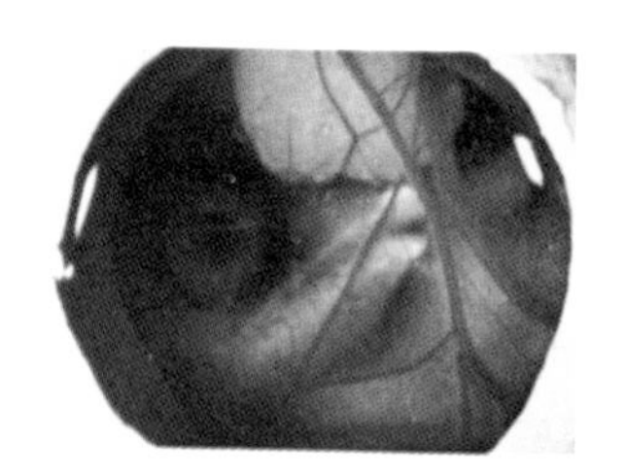

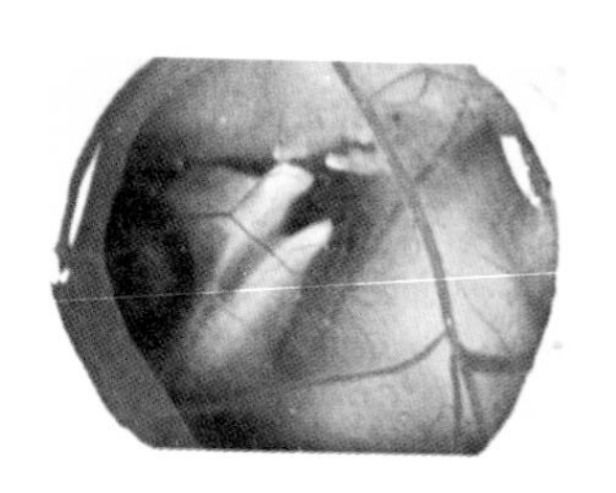

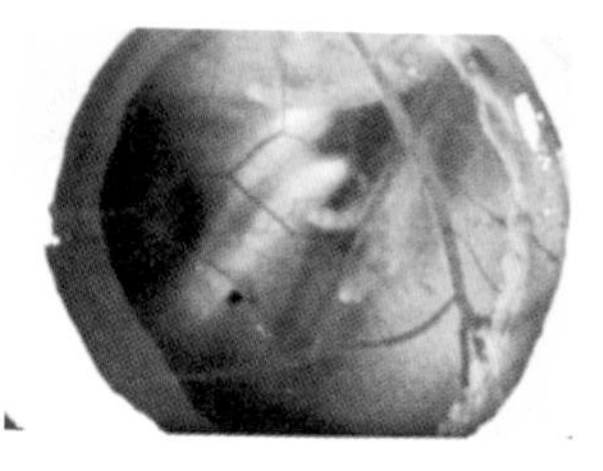

bird drinking, you might recognize that same quick opening and closing, like a shivering of the beak, that accompanies swallowing. Doing this in the egg might help the chick to be prepared for next week, when it will be out in the world.

This week the chick can no longer lie across the middle of the egg as before. The chick has become too big. By wriggling about, it gradually comes to lie lengthwise in the egg with its head in the blunter half and the tail in the narrower one. This chick has nearly outgrown its first home, and it must begin to move out.

When the moving day approaches there is always about a thimbleful of yolk and a bit of albumen left in the egg. The chick will take them along. Of course it cannot pick up and carry them. Instead, a most remarkable thing happens on the eighteenth or nineteenth day in the egg. The whole leftover yolk, enclosed in the yolk sac, and the leftover albumen in its sac, are taken into the chick's belly. They are slowly drawn into the body through the opening at the navel. This gives the chick a good bellyful to live on for its first day or two out in the world, so that it can rest and sleep before having to hunt for food.

However, there is one thing the chick must be ready to do instantly and regularly as soon as it leaves the egg. This is breathing. As you will remember, the chick in the egg receives its air, or oxygen, directly in its bloodstream through the allantois. In the last two days in the egg, while still getting oxygen delivered in this way, the chick also gets a chance to test its lungs and to

1. Head up, opens beak 2. Head down, closes beak 3. Scratches beak 4. Moves toes

rehearse breathing. A day or two before hatching, the chick will poke its beak up into the special air space under the eggshell, at the wider end of the egg. Do you remember the small air-filled space between the shell membrane and the shell of your kitchen egg? This is the air space which the chick is now going to use.

The chick seems to have been preparing for this ever since its first week in the egg. As you will remember, it has been repeatedly lifting up its head since then. Finally, by the nineteenth or twentieth day, the chick is so big and its beak is so strong, that one more lift of its head, and the beak pops through the amnion balloon. It pokes through the amnion, and then it pokes right on through the shell membrane overhead and reaches into the air space. The chick's two nose openings are on the beak, so the chick's nose reaches the air space, and the chick can take a good breath of air for the first time. Since the chick, at the same time, is still getting its supply of oxygen from the allantois in the egg, it does not yet need to breathe regularly. It is merely trying out its lungs.

With the first breaths of air-space air, the chick is also able to make its first noises, loud cheeping noises. You cannot yet see any break in the eggshell, but when you hear that cheerful, loud sound coming from the inside of an egg, you know there is a chick, nearly ready to arrive. If you clap your hands or make any sharp noise near the egg, the chick will answer with a loud cheep from inside. You can expect to see that chick in a day or two.

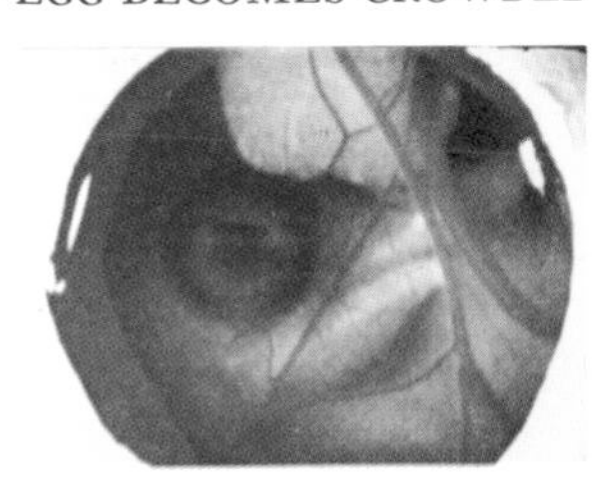
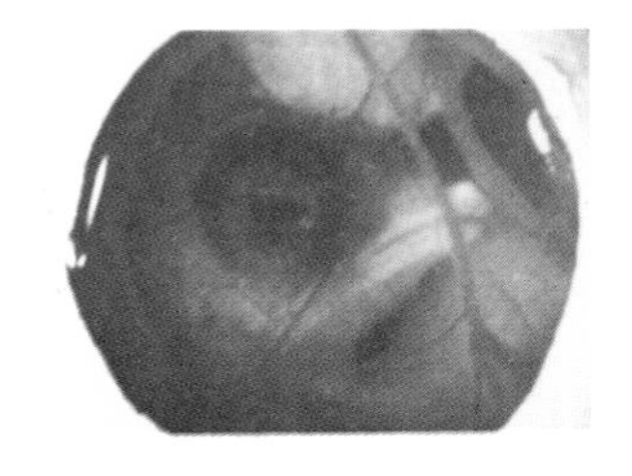
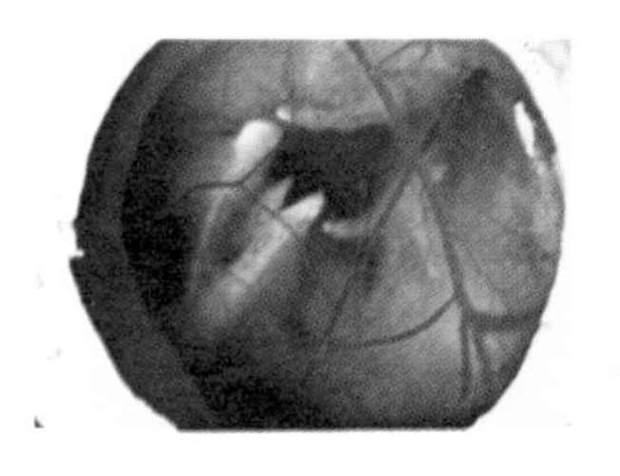
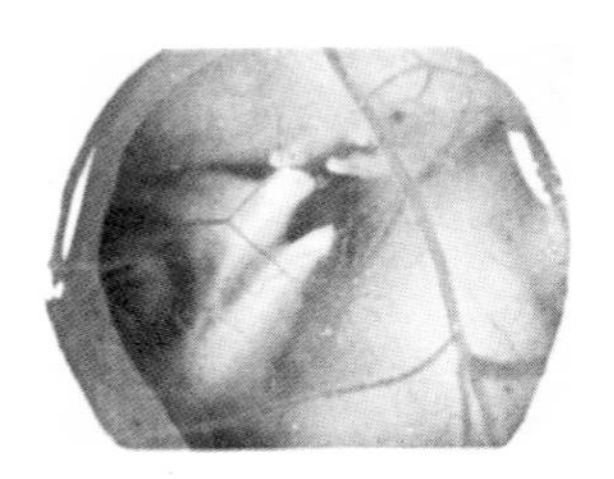

1. Takes nap 2. Half opens eyes 3. Opens beak wide, shows tongue 4. "Toes in my face again!"

DAY
21

Hello, Chick!

Enough of being shut up inside an eggshell! After twenty-one, sometimes twenty-two, days in the egg, the chick makes a window in the shell and pokes its nose out into the world.

This chick is pipping. What is pipping? It is the special word for what the chick does when it cracks the shell with its beak. While pecking through the shell like that, it usually also peeps—it makes loud noises. It is a pipping, peeping, pecking chick. It is a hatching chick, fighting its way out of the egg, ready to be born.

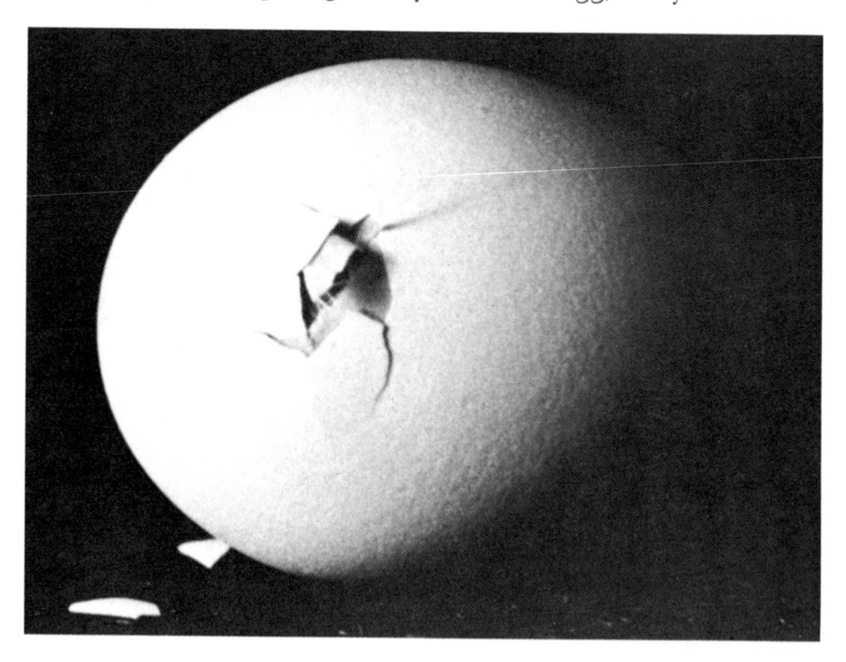

Pipping chick . . .

. . . pokes its nose out. "Pimple" near point of beak is special pipping tooth.

Hatching is a struggle for our chick, and it will take several hours of hard and exhausting work. The chick gets so tired that it works and then sleeps, and works and sleeps, until it is finally free of the shell.

It is easy enough to break an egg when you drop it on the floor. But imagine yourself stuffed tightly into three sacks, all made of stretchy, rubbery material, and holding you down so that you can barely move. That is how the chick's body is now tightly enclosed by its surrounding balloon and by the two shell membranes. Only the beak is poking out. On top of these three sacks imagine a hard shell which, to a small chick, would be about as tough as an egg made of thick cardboard might be to you.

The chick could not get out at all if its beak had not grown so large and hard in the last week, or if the chick had not developed the ability to raise its head and bring the beak up sharply. As you will remember, two days ago the chick, in raising the head like that, poked its beak through the membranes, and the beak reached the air space. Then the chick raised its head again and again until, two days later, the beak happens to make the first crack in the shell. In good time for this important occasion, special muscle reinforcements have grown at the back of the chick's neck. These are called hatching muscles, and they will disappear soon after hatching. Also for the occasion, a small hard bump has grown on the top side of the beak. It is called the pipping tooth. It is useful for breaking away bits of shell; it has no other use in the life of the chick, and a few days after hatching it will drop off.

With beak and pipping tooth and hatching muscles, and with

One final . . .

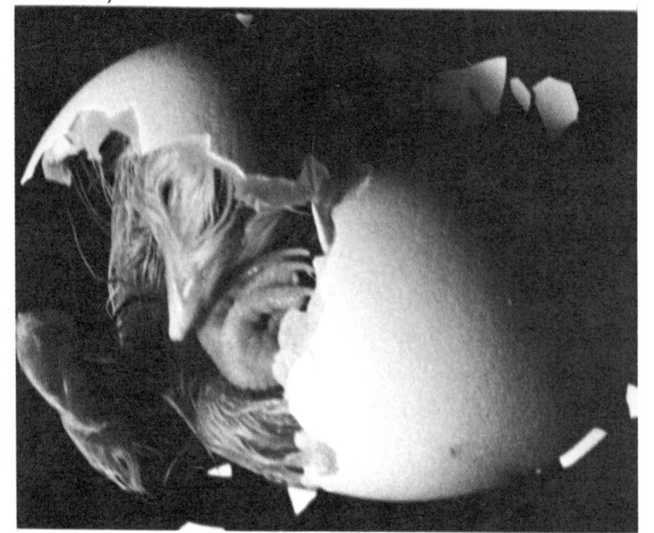

Great effort . . .

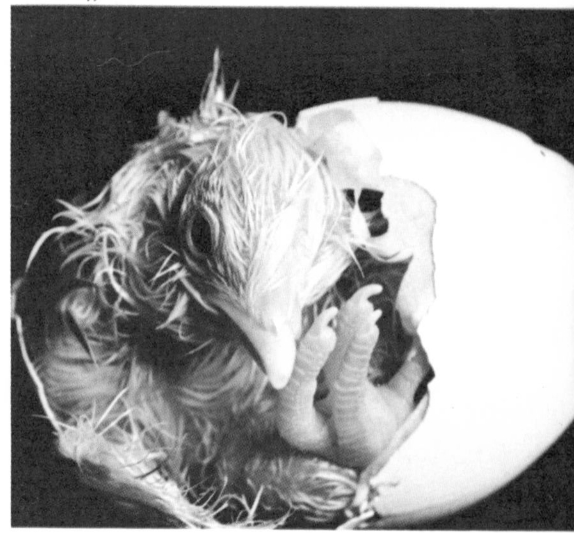

great effort, the chick finally manages to make that small hole in the shell. Having accomplished this, it needs a rest and is content to lie by its new window for a while and have a nap. Then it wakes up refreshed and all excited. Cheeping loudly, it begins to try to free its head from the shell membranes. To do this, the chick turns the head as much as possible, which is not very much, from side to side. This also cracks bits of shell around the new opening, making the eggshell window larger bit by bit. It goes very, very slowly. An hour later the opening is only slightly larger. A few more naps, many more efforts, and another two or three or even ten hours go by. The chick keeps on trying to stretch out in the egg. It braces its big feet against the inner wall of the eggshell and pushes and wriggles until, at last, the whole eggshell begins to show a crack across the middle. Then more heaving and pushing and wriggling as the egg gradually breaks in half. Ahhh — the chick can finally stretch. The chick is out!

Nearly . . .

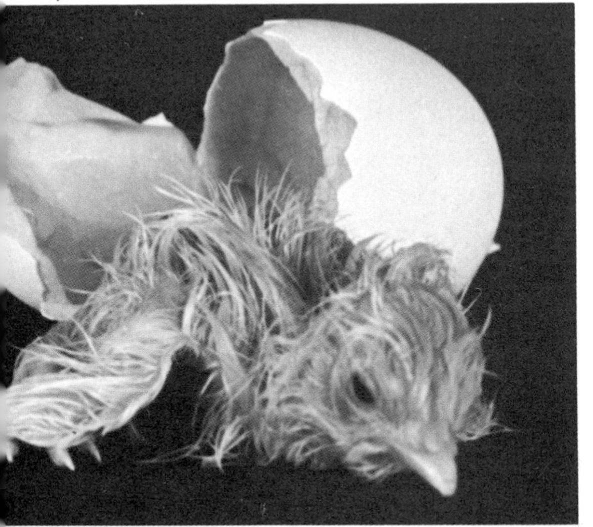

Out!

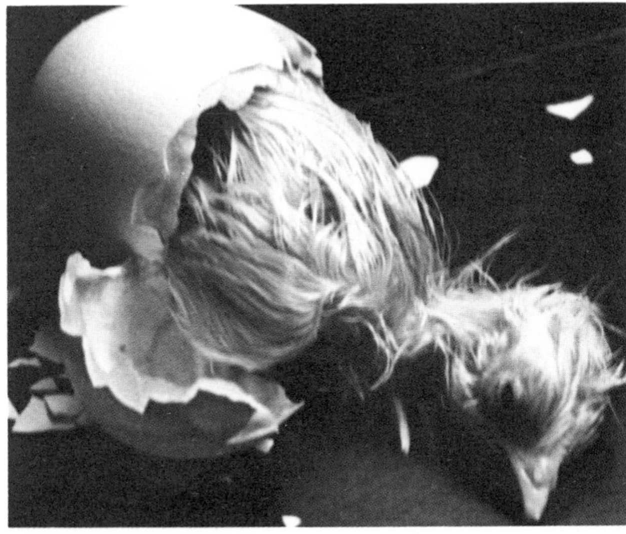

What a disappointing arrival. The chick that has been cheeping so loudly, sounding so proud of itself, looks awfully bedraggled as it flops, rather than hops, out of the egg. It will take another day for this chick to be really fit for its first public appearance.

Our hatched chick is wet and tired, and it is stunned by many new experiences: open space, brighter light, fresh air, the need to breathe without stopping. How different from the warm, close protection of the egg, where all its needs were cared for through the special blood vessels and membranes that are now left behind in the eggshell.

The wet, downy feathers cling to the body. They are not wet with sticky yolk, they are wet from the waters in which the chick has been living until now. The chick shares this wetness at birth with all other animals; all arrive clean and wet because, as you already know, they must grow protected in fluids. Fish are wet in the ocean, tadpoles are wet in a pond, all land animals are wet from the tiny pool of waters in which they have grown, either inside an egg or inside their mothers, depending on who they are. You also were wet at birth.

Wet, cheeping chick squats on big feet, not yet able to stand properly.

Tired chick cuddles up to a companion in incubator.

The time for birth, or hatching, comes when the growing young one must move out, because there is no longer enough food or enough space for it. At that time the young one is ready to start a new way of living. Many, like you, will still need a lot of help from their parents. You were born quite helpless. Others, like the chick, have to be more ready to look after themselves.

The chick quickly has to learn to stand on its own. It has good big feet, but on this first day the legs are still very wobbly and cannot carry the chick far from its egg. Right now this chick is expert only at cheeping, which it does very loudly, and sleeping, which it will soon settle down to do.

As you know, the chick has enough yolk tucked into its belly to be able to wait a day or two before looking for food. On this day of hatching, the only thing that must be provided for the chick is plenty of warmth. In an incubator, the chick must be kept as warm as it was in the egg for the next several days. In the farmyard it will cuddle under the mother hen. Right from hatching, chicks seem to have the sense to complain loudly if they are chilly, and they know to settle down only in a warm place, where the downy feathers will soon fluff out. Then the chick will be really ready to explore its new world.

Good-bye, Chick

DAY 22 DAY 23

Twenty-two days ago this was one cell, so small you could not see it without a microscope. Today it is millions of cells that spell a cocky chick.

What the cells have built is much more than the form of a fluffy yellow chick. It is a chick with a brain that can receive and send messages, with nerves that can carry the messages, a heart that beats and can pump blood, a stomach that can digest food, lungs that can work in breathing, eyes that can see, ears that can hear, a nose that can smell, legs that can walk, claws that can scratch, wings that can flap, a beak that can open and close, bones and muscles for strength, feathers that will keep it warm, and hundreds, even thousands of other parts that move and work.

But the cells have built still much more than that. They have not only made a chick that is able to do many things, they have made a chick that seems to know when and how to do them. Nobody has to teach the chick how to eat, how to walk, how to drink, how to find food when it is hungry, how to preen its feathers, how to act with other chicks, when to run away from danger, to stay away from places where it might be too cold or too hot, never to stray too far from its mother, and to huddle together with other chicks for warmth and safety.

Right from the start the chick is a busybody taking charge of

Pecking for food

Making friends

Checking in with the family

"Don't go away!"

its own life as if it knew all about the world. This is very different from some other animals, and also from people. People are born with little know-how but with much better brains, and with a far greater ability to learn. The chick will learn, too, to some extent. But much of its everyday behavior will always be ruled by the "message" from the first cell in the egg. This includes all the chick needs to know when it first comes out of the egg.

Putting its abilities into action, one day after hatching, our chick here struts off to a new adventure. It is on its way to becoming a chicken. In the next few weeks it will grow larger, its downy feathers will be replaced by larger quill feathers, and in about half a year this chick may itself become a parent. Good-bye, chick.

Good-bye.

Index